Living the Lighting Life

Living the Lighting Life provides practical tools and advice for a successful career in entertainment lighting.

This easy-to-navigate guide offers real-world examples and documentation from the author and key industry experts, giving readers a comprehensive overview of the lighting life. The book provides insight on:

- Different job opportunities in the entertainment lighting industry;
- Business procedures, contracts, time sheets, and invoices;
- Tips on self-promotion, networking, and continual learning;
- The lighting lifestyle, healthy living, and work-related travel;
- Maintaining and developing creativity to provide innovative lighting and solutions.

With insightful interviews from industry veterans, *Living the Lighting Life* is a key navigational resource for anyone considering a career in entertainment lighting or just starting out.

Brad Schiller has over thirty years of experience in the entertainment lighting field. A self-described "lighting geek," his deep-rooted passion for lighting began at a very early age. Brad's career has allowed him countless opportunities to work around the globe, and across diverse facets of the industry. He is also the author of *The Automated Lighting Programmer's Handbook*, now in its third edition (Routledge, 2016).

Living the Lighting Life

A Guide to a Career in Entertainment Lighting

Brad Schiller

Routledge
Taylor & Francis Group
NEW YORK AND LONDON

First published 2021
by Routledge
605 Third Avenue, New York, NY 10017

and by Routledge
2 Park Square, Milton Park, Abingdon, Oxon, OX14 4RN

Routledge is an imprint of the Taylor & Francis Group, an informa business

Library of Congress Cataloging-in-Publication Data
Names: Schiller, Brad, author.
Title: Living the lighting life : a guide to a career in entertainment lighting / Brad Schiller.
Description: New York, NY : Routledge, 2021. | Includes index.
Identifiers: LCCN 2020016988 (print) | LCCN 2020016989 (ebook) | ISBN 9780367349332 (hardback) | ISBN 9780367349325 (paperback) | ISBN 9780429328787 (ebook)
Subjects: LCSH: Stage lighting–Vocational guidance. | Stage lighting designers–Interviews.
Classification: LCC PN2091.E4 S35 2021 (print) | LCC PN2091.E4 (ebook) | DDC 792.02/5023–dc23
LC record available at https://lccn.loc.gov/2020016988
LC ebook record available at https://lccn.loc.gov/2020016989

ISBN: 978-0-367-34933-2 (hbk)
ISBN: 978-0-367-34932-5 (pbk)
ISBN: 978-0-429-32878-7 (ebk)

Typeset in Bembo
by Swales & Willis, Exeter, Devon, UK

This book is dedicated to the entire entertainment lighting industry and all its players.

Contents

Figures

Tables

Introduction

When I was about twelve years old, I was very much into magic. I actually performed at school events, parties, and more. During this time, I thought that I wanted to be an actor, but instead I discovered my passion for the technical side of the industry. Learning the ins and outs of the illusions provided me with a background for what would become a fantastic career.

My parents were very accommodating in most of my endeavors growing up, and they supported me in the majority of things in which I had an interest. After junior high, they searched for a private school to send me so to that I could have a stronger education with better opportunities for college. Little did they know how things would actually turn out. The school we selected was based on both its college preparatory status as well as the theater program. Remember, I was interested in being an actor. After signing up for my first acting class, I learned that I must also learn about technical theater. This is where I met Mrs. Cheryl Ellis.

This wonderful woman taught me many backstage skills, including learning about lighting. I quickly found a calling to focus on lighting and special effects. In fact, I was only an actor in one scene of my first play at the school; the rest of that show I was at the light board running cues! For the next three years, I was very involved with hanging and focusing lights, designing plots, running cable, and all the general crew work I could get my hands on. I would even skip my English classes to spend more time in the theater. Actually, though, that did not work out so well as I soon found I was failing English and told I could either leave the school or continue without theater. My parents supported my decision to leave the school.

When I entered the local public school in my junior year of high school, I was excited to move forward with my technical theater training. However, when I asked the theater teacher about the technical theater offerings, she said, "You mean the things that the people who don't get cast in the show do?" I knew right away that this program was not for me! Luckily, I discovered an amazing opportunity right across the street.

The Irving Center for Cultural Arts was an old converted movie theater that had a permanent set on the stage and a well stocked lighting inventory. I immediately began volunteering my time to community theater groups as a lighting technician. I was thrilled to be back in the theater and growing my knowledge of lighting. Here I was working alongside both amateurs and professionals who were happy to share their knowledge with me. Occasionally the house staff would pay me to run a followspot or help with a focus.

Within a short matter of time, some of the permanent staff were moving on and I was offered a part-time position. After a few years, the other staff left, and I found myself running the entire facility! I was booking events, hiring my friends, designing shows, focusing lights, selling concessions, and everything else I could get my hands into. I was an official city employee and was making more than minimum wage while attending classes at a local community college.

At the same time, some of my friends went to work at this place in Dallas called Vari-Lite and they asked if I wanted to join too. I learned that it involved being on the road touring with bands, and I decided that I would rather stay and concentrate on theater.

I also started working with the IATSE union and a few local comedy troupes, to further expand my career. I was hungry for gigs and took on as much as possible. I was fully engrossed in lighting at this point and knew it was where I wanted to be. I would read trade magazines, take classes, study the credits on television shows, and learn from those with more experience. I also attended the first-ever LDI tradeshow, which happened to be held in the same city where I lived and worked.

At that first LDI show, I came across a booth filled with Intellabeams and I learned all about High End Systems. I met Richard Belliveau and Tim Grivas, which formed relationships that still continue to this day. I had a great interest in the new field of moving lights and looked for any opportunity to get my hands on them. I started programming with the earliest of controllers and taught myself all that I could. Eventually I was able to attend one of the first-ever programming seminars held by High End Systems. While continuing to work at the Irving Center of Cultural Arts, I would freelance around town as a tech or programmer.

Over time, the City of Irving built a new state-of-the-art complex called the Irving Arts Center. I took on a new position as technical director, in charge of the smaller of two theaters called the Dupree Theater. I was also able to attend more LDI shows and take part in industry learning on a greater scale. I never stopped freelancing as a technician, programmer, and designer. Around this time, I met Eric Durr, who owned a small number of Intellabeams and he presented me with a tremendous opportunity to learn about moving light programming. My life was focused around lighting, and yet I had bigger goals to get accepted into California Institute of the Arts so I could earn a degree in lighting design.

In 1995 I packed my things and headed out to California. I was thrilled to start as a junior at CalArts in their lighting design program. While I enjoyed the other students and the opportunities, I was often asked why I was at the school. At this point I had over eight years of professional work in the industry. I had far more experience programming automated lighting than anyone at the school. While in school, I was also freelancing around the Los Angeles area as much as possible. After one semester, I decided to leave the school to further pursue my professional lighting career.

I started working at Towards 2000 in Burbank, CA under Mark Rowlands. I told him that I wanted to be a full time programmer, but he instead offered me a job as a rental manager with the option to program when possible. I would answer the phone and send out rentals during the day, while going on gigs nights and weekends. I never stopped learning and kept reading, studying, and meeting all the professionals that I could.

During this time, I was introduced to a brand new lighting controller called the Wholehog II. I quickly latched onto it and started to email with the creators of the desk. Within no time, I was one of the only Wholehog II programmers in Los Angeles, let alone the country. I was not just programming, but also beta testing, writing libraries, requesting features, and more.

Early in 1996, Towards 2000 had the opportunity to work with Martin Lighting to provide the first run of Martin PALs for that year's Academy Awards Ceremony. I was thrilled to be a part of this mega event as a crew chief, as I had dreams of one day designing the show. Then when the prototype Case console Martin was using failed during pre-production, I was presented with an even better opportunity than crew chief. On the Friday before the Sunday night live broadcast, I fired up my Wholehog II running v0.5 beta software and began programming the PAL fixtures for the Oscars. It was a thrilling experience that certainly checked a major line in my bucket list.

Coincidently, during my time at this show, I was contacted by Tim Grivas at High End Systems. He offered me a job to move to Austin to be a part of the lighting programming team. I jumped at this chance, and a few months later moved to Austin to start my new role.

Over the next seven years, I programmed countless productions including concerts, corporate events, theatrical shows, television productions, and more. In 2000, I was part of a team of seven programmers on the Olympic Games Opening and Closing Ceremonies in Sydney, Australia (see my diary of this time period in *The Automated Lighting Programmer's Handbook*). While working at High End Systems, I also learned about the processes of fixture and console development, and assisted in the launch of many new products.

My original plan was to work for High End Systems for five years and then return to a freelance design and programming position. However, new products and major economic changes as a result of 9/11 delayed my transition. After seven and a half years, I finally made the jump back to the world of freelancing. Once again, I could never have predicted how my career would transpire from this point.

About three months into my new entrepreneur endeavor, I received a phone call from John Broderick, who was the long time LD for Metallica. We had previously worked together on many ice skating shows and he needed a programmer/operator for the upcoming Metallica world tour.

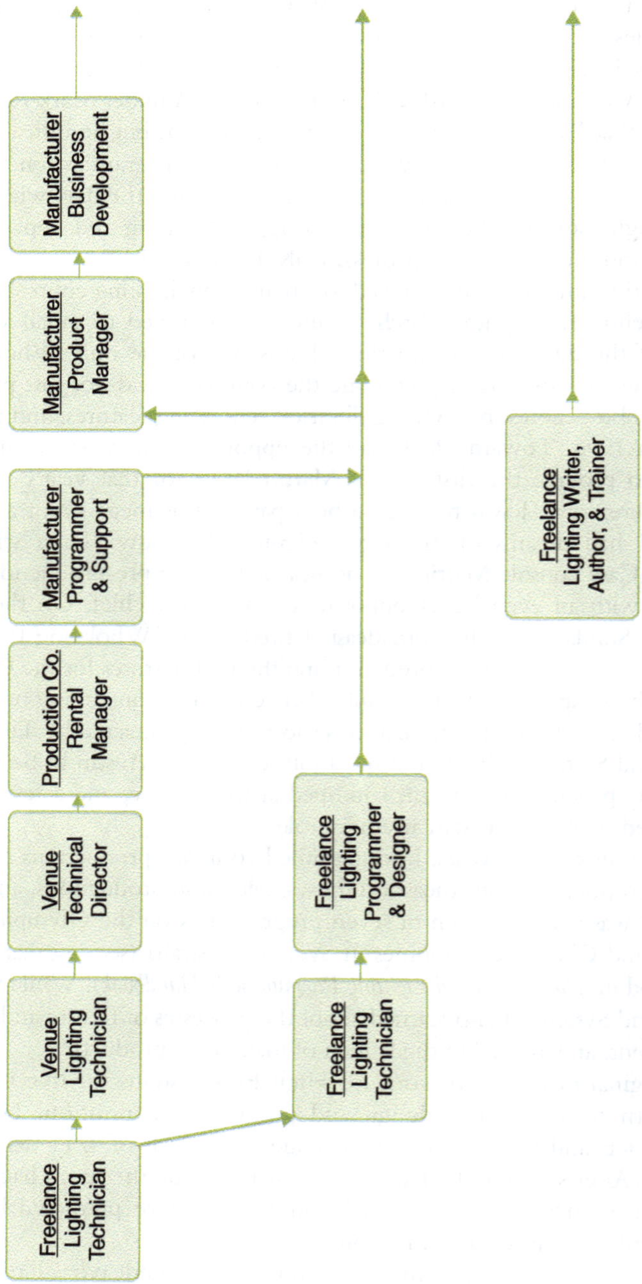

Figure 0.1 Brad Schiller's Career Path

After a little negotiating, I found myself on a tour bus running Metallica shows alongside Butch Allen. We travelled the world for about a year and half; I also managed to fit in a few other freelance gigs along the way.

Then, while working the Ultra Music Festival in Miami, Richard Belliveau called and suggested that I come off the road and back to High End Systems. I had never actually considered going back, but he offered me the chance to take the lead on developing new lighting consoles (which eventually became the Hog4). So I took the job and worked with an amazing team developing many new Wholehog consoles (Hog IPC, Road Hog, Full Boar) and restoring the Wholehog brand to a prime place in the market. Plus I continued to freelance a bit on the side, programming concert tours, festivals, corporate events, and television shows.

A few years later, High End Systems was sold to Barco and my position changed to include not just console product management, but also fixtures. In my wildest dreams as a young theater technician, I never knew that I could have an impact on the development of new lighting and console products in our industry. Working with a wonderful team of engineers, we developed many fixtures that continue to share innovations within the industry.

After another seven years at High End, I decided it was time to move on, and I took a new position at Philips as product manager for their Vari-Lite and Showline product lines. Here I worked with different teams of engineers based in Dallas and China to specify and help develop new products. As LEDs began to infiltrate the industry, I found it particularly interesting to work within this new medium.

Everything changes within big corporations such as Philips, and I found that the classic Vari-Lite that I believed in was unfortunately no longer alive. I decided to leave after two years of employment, and at that time I started working with Martin as a business development manager. For the past five years, I have been the company's liaison to designers and programmers, and able to focus more on the creative aspects of the business versus product development.

Throughout my career, I have dedicated myself to remaining involved with productions and continuing programming and/or designing whenever possible. I have been fortunate to have maintained a steady paycheck for most of my career; only at a few points was I a full time freelancer. This is one of the things I love about the entertainment lighting field. There is a plethora of opportunities and you never know what is around the corner. As long as you keep doing what you love, you too can have a satisfying and profitable career in the entertainment lighting industry.

My wish for anyone reading this book is that you craft your own story with diversity and multiple opportunities throughout your career. Don't be afraid to fail or change directions, and know that prospects for success are abundant yet may not always follow your plan.

In the following pages I will present you with guidance and information to help you through your career. Much of the text is general by design as

I am not in a position to provide specific legal or financial advice. However, I have consulted many in the industry to ensure that various diverse options and resources are made available to you. No matter where you are in your career, I trust that you will find some nuggets of wisdom to help you along your path.

Have fun, be safe, and may you have the same love and passion for lighting as I do.

Part I

The Job

The entertainment lighting field is a large industry with many different positions that one can hold throughout a career. Productions occur across a broad range of genres and allow our work to be seen in a variety of settings including concerts, theater, television, film, corporate events, night clubs, houses of worship, and more. Understanding your role regardless of genre is a key element of a successful career.

1 Lighting Jobs and Roles

The entertainment lighting industry consists of a plethora of vital positions that each come with their own set of unique qualities and abilities. You can make a career within just one of these roles, or you can progress through many. No matter your path, it is important to understand the jobs that are available within our industry.

Production Lighting Positions

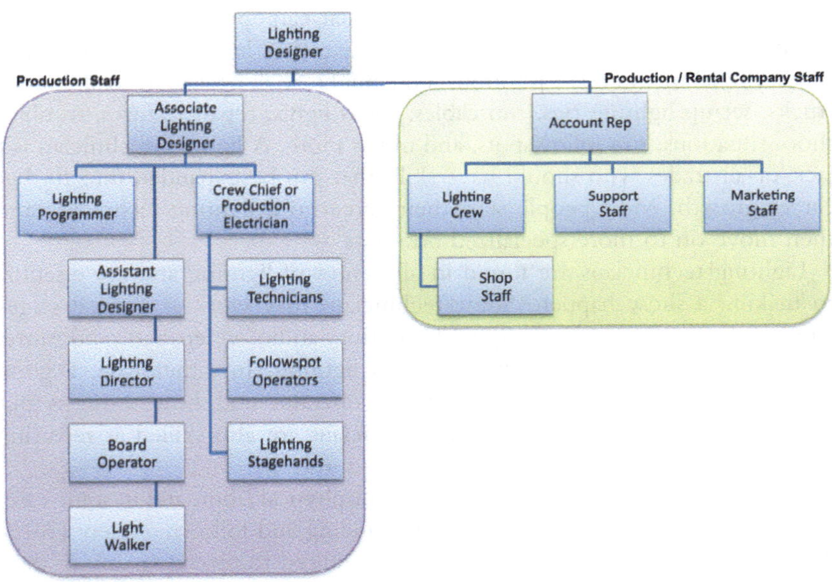

Figure 1.1 Production Lighting Positions

Shop Technician

The shop technician generally performs his or her tasks within the confines of a lighting or production shop. Specific duties range from warehouse personnel, to show prep, and fixture maintenance. In many cases, people begin their careers working in a lighting or production company where they can hone their skills, gain equipment knowledge, and understand systems before participating on a crew for a show.

Show prep duties allow shop staff to work alongside production professionals as they lay out trusses, prepare cable runs, test fixtures, and sometimes even begin programming. This provides a great environment to gain valuable experience and learn from others. Furthermore, common show practices such as cabling, safety rigging, labeling, and other procedures are standard processes during shop prep.

Others will choose to specialize in fixture repair and work within a lighting shop to maintain the equipment and ensure it is ready for each production. Shop technicians are an integral part of the industry as their preparation, repair, and quality control abilities have a great effect on the outcome of a production's lighting.

Lighting Technician/Stagehand

Lighting technician is a very broad term that can apply to many different aspects of production and personnel. In general, a lighting technician is a physically labor intensive position. You might be expected to load/unload trucks, set up lighting rigs, run cables, focus lights, repair fixtures, troubleshoot situations, run followspots, and much more. A lighting technician is a jack-of-all-trades who should have skills that can be expanded into further career growth. Most people start their career as a lighting technician and then move on to more specialized positions.

Lighting technicians are found in all genres of lighting and are essential to making a show happen. Some technicians may work as general stagehands, where they are working with lighting, audio, video, sets, and more. Others will segment themselves only to lighting. In either case, a good knowledge of how lighting equipment operates and how a rig is put together is essential. Some lighting technicians are also skilled at repairing products, but this is not a requirement for most.

A lighting technician must be prepared for physical labor, and in some cases climbing and working at heights. Understanding and following safety procedures is a must! Technicians generally are required to provide many of their own tools, and follow the guidance of the crew chief or master electrician.

Crew Chief/Production Electrician/Gaffer

A lighting crew chief, production electrician, or gaffer is in charge of the entire lighting crew as well as overseeing all aspects of the lighting rig. This role is managerial, as the crew chief must oversee all matters relating to the lighting technicians and other crew members. The crew chief is the go-to person for all lighting matters. This position has great responsibility as the entire lighting rig operation falls back to this person.

Furthermore, the crew chief, production electrician, or gaffer will often be involved with drawing lighting plots, determining shop orders, interacting with the vendors, and helping the lighting designer achieve the lighting vision. Typically in touring the position is referred to as crew chief, in corporate and theater as production electrician, and in television and film as gaffer.

Master Electrician

In some cases, the term "master electrician" is used interchangeably with crew chief or production electrician. However, in other circles, a master electrician is responsible for the actual electricity portion of a lighting rig. This person will come up with the proper cable and electrical distribution plan, as well as connect the power at the venue. The master electrician will measure and monitor the power throughout the production and ensure that all electrical components and connections are correct and safe.

Lighting Programmer

Nearly every lighting production makes use of a lighting console and the person that inputs the data into the desk is known as the lighting programmer. This job requires both creative and technical skills, as the programmer must understand the syntax and process of programming as well as the nuances and procedures of lighting. With the advent of automated lighting fixtures in the 1990s, the programmer has become very much a creative assistant to most lighting designers.

A programmer will often share the duties of creating looks, choosing colors, and determining focuses with the designer. Of course, the LD has the final say, but because the programmer understands what can be accomplished on the console, his or her input is often invaluable to the LD. Full details of the responsibilities and tasks of a lighting programmer can be found in my book, *The Automated Lighting Programmer's Handbook.*

Lighting Director/Lighting Board Operator

The lighting director position originally came from the television world and described a position similar to today's lighting designer. In some cases, the television term of lighting director still relates to the person in charge of the overall lighting of the production.

However, in most other cases the lighting director is a person on a touring or corporate production who operates the lighting console and also calls followspot cues. Thus, they are directing the lighting that was initially put into place by the lighting designer and programmer. A lighting director will also understand basic console programming and update focus positions and cue timings as needed. Furthermore, the lighting director is responsible for maintaining the look of the lighting throughout the tour. Often this person will make daily decisions about the lighting rig, cues, or even create new cues for songs or scenes that were not previously designed and programmed.

In many theatrical and house of worship genres the position holds less responsibility and is known simply as a lighting board operator. Usually a stage manager will call the cues both to the lighting board operator and the followspots. The lighting board operator does not change the programming or have any responsibilities to contribute to the look of the lighting. A board operator merely plays back the cues as instructed.

Lighting Designer

Often considered the pinnacle position in the industry, the lighting designer (LD) is the ultimate creative person regarding the lighting of a production. This highly regarded position must interact with everyone on the production from the top down. From the producers, to the artists, to the lighting crew, the LD is responsible for the lighting on the production. From the initial concepts, to the light plot, to the selection of colors, chases, and other lighting looks, the buck stops with the LD.

A lighting designer must understand not only how to light a show, but also how the lighting rig is put together, how it is transported, the budget allowed, the associated costs, and much more. The level of responsibility is highest with the lighting designer, and thus why this position carries more weight than all the others. Of course, with the highest position usually comes the highest pay, as well as the greatest levels of stress. It has been said that often more of a lighting designer's duties are political than actual lighting.

Some lighting designers take on multiple positions such as lighting director and/or lighting programmer. In some smaller markets, the lighting designer may be a jack-of-all-trades and even be a crew chief and/or technician on the same production. The lighting designer is also a liaison with the rest of the production team, artists, directors, band members, management, and more. Many LDs will rely on a team of additional lighting people to help them achieve the lighting vision while the LD is busy meeting with all the other stakeholders in the production.

Assistant/Associate Lighting Designer

The role of an assistant or associate lighting designer is to work alongside the lighting designer and help as needed. Duties may include drawing plots, providing lighting look input, creating and organizing paperwork, fetching coffee, attending meetings, choosing colors, tracking data, and more. Depending upon the assistant/associate lighting designer's skill set and relationship with the LD, the tasks can run the gamut from the mundane to the highly creative.

Assistant/associate lighting designers hold a coveted position as they often get to work alongside the lighting designer, picking up valuable knowledge, resources, and contacts along the way. Typically an assistant is the lower rung of these two positions, whereas the associate is a closer collaborative to the lighting designer. The associate will contribute more to the look of the lighting and have greater responsibilities than the assistant. However, depending upon the genre, the terms may be used interchangeably. Typically in theater the roles and labels are well defined, but not so much in touring, corporate, and house of worship environments.

A great resource for learning the duties of an assistant lighting designer is Anne E. McMills' *The Assistant Lighting Designer's Toolkit.*

Light Walker

A light walker is someone who is a stand-in for actors, allowing the LD and crew to focus and view lighting looks when the actors are not available. Often a light walker will stand for hours on the stage with light pointing directing into his or her eyes. In many cases they must also hold their hand(s) up to approximate the height of an actor. While not a glamorous or skilled position, it can be very beneficial to some in the industry.

If you are able to be a light walker for a well-respected LD, then this can give you a great insight into their work. You will get to see the LD in action and also possibly interact directly with the LD and other lighting staff. While not all light walkers get paid, the opportunity and contacts you make could be very beneficial to your career.

Instructor/Educator

As with any field, there is a strong base of entertainment lighting professionals who choose to teach and share their craft with others. Many different educational programs including high schools, colleges, private schools, and independent educators serve an extremely useful function within our industry. Generally the instructors have a vast amount of real production experience, and most are also still working with productions while also teaching.

Account Sales

An often overlooked position within the entertainment lighting industry is that of sales. There are two key genres of salespersons within the field. First, lighting or production company salespersons work directly with lighting designers to help them achieve their vision, utilizing the equipment on hand and within the specified budget. Quite often they are referred to as "account reps." Then there are salespeople who work for manufacturers or distributors who negotiate deals with lighting and production companies to ensure that they are purchasing the latest equipment on the market (see Lighting Manufacturer Positions).

The rental or production company salesperson must have a good working knowledge of production, as this person will often make suggestions regarding gear, trucking, staging, crew, and more to assist the LD before a production leaves the shop. Most account reps have a production background, having worked as technicians, crew chiefs, or LDs on many shows.

Every LD has their favorite salesperson at various lighting shops and this relationship-based position is key to productions coming in on budget and on time. A good rental or production sales person will also be the production's point person regarding all the lighting equipment and staff. Often the salesperson will be onsite during pre-production to ensure everything moves smoothly and to assist the lighting team with any last minute needs they many have.

Lighting Manufacturer Positions

Lighting equipment manufacturers are large businesses employing many people across various disciplines. There are multitudes of ways a person can become a part of the entertainment lighting industry while working with a lighting equipment manufacturer. All the typical positions held within any company are available (HR, accounting, legal), but only a few positions really specialize when talking about the entertainment lighting industry.

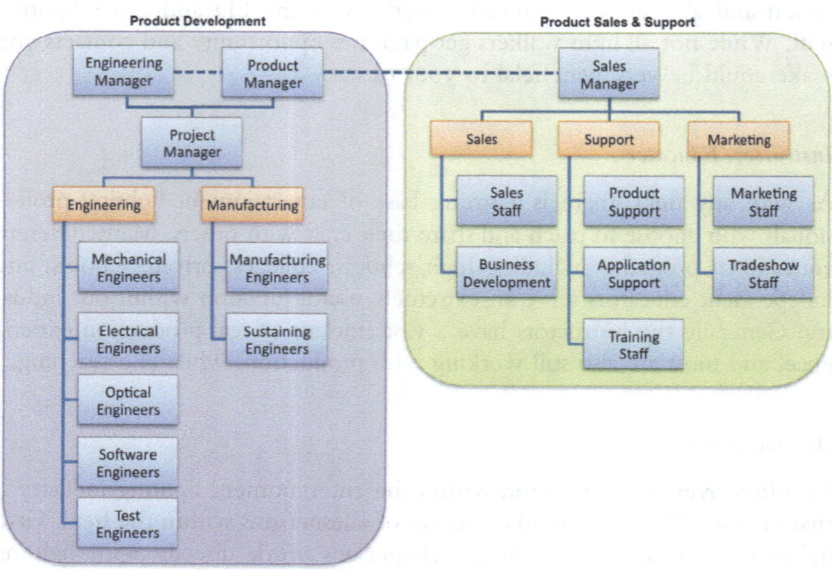

Figure 1.2 Lighting Manufacturer Positions

Product Development

Electrical, mechanical, and optical engineers, project managers, computer programmers, product managers, and more are required to create and develop new lighting products. Many of these positions are often filled by staff who have never been backstage at a show, or even have a full understanding of what it takes to put up a lighting rig. However, their specific engineering backgrounds enable them to work within our field to create exciting new products.

Equipment Support

Lighting equipment manufacturers must have a strong support team to assist productions when units fail, and to train the industry on the operation and repair of their products. Technical support staff will be experts with their particular products, and should be able to describe solutions to nearly any problem that develops. In addition, they often teach training classes to those working in the industry. Some courses explain how to program consoles, media servers, and complex fixtures, while others are deeply technical repair programs.

Many support personnel working for manufacturers have a background with numerous years of actual lighting productions in various positions. Their real-world knowledge added to their product specific familiarity makes them extremely valuable resources for anyone using the specific equipment. After several years, many of these product experts move on from the manufacturers to become freelance technicians, programmers, and lighting designers.

Application Support

An offshoot from typical product support, an application support department will help productions directly by providing manufacturer staff to assist with various aspects of a lighting production. From assisting with fixture layout and usage, to understating components and accessories to order, they provide an invaluable resource for customers. Often these experts will be onsite at pre-production and work alongside the show's staff to properly utilize the manufacturer's equipment as required by the production. Similar to traditional support personnel, application support staff members usually have a strong background working within the lighting field.

Business Development

Business development roles are typically held by industry veterans who are experts across many genres of the field. They will have a strong relationship with a vast number of lighting professionals, as well as a strong understanding of production processes and procedures. Their core purpose is to be a liaison between the lighting designers, programmers, technicians, and the lighting manufacturer.

The business development staff members will also ensure that those making decisions about products in the industry are aware of the products offered by their particular manufacturer. Furthermore, they will work with the sales team, securing specifications of specific products and increasing product demand leading to increased sales.

Sales

Every equipment manufacturer and distributor has a sales team that is tasked with selling new product into our field. Because most shows do not purchase their own gear, these salespeople must interact with the lighting rental and production companies. Many of the manufacturers' sales staff may not have worked on actual productions, but instead have a strong background and training in sales.

However, they must quickly learn the procedures and particularities of our industry in order to properly sell their wares. Furthermore, they must be experts with their company's particular equipment and understand the needs of the LDs and other lighting staff.

Marketing

Every manufacturer needs to have a strong marketing department to assist in promoting the products and their usage. Many will hire industry experts who have a strong marketing background as well as a vast "black book" of lighting contacts. These specialized marketing people will work directly with lighting designers, programmers, technicians, and others to promote their work that utilizes the manufacturer's products.

Furthermore, they will often assist in industry tradeshows and need to have an understanding as to how lighting productions operate. Many will form strong relationships with programmers, lighting designers, and production or lighting company staff.

Many More Opportunities

There are countless more opportunities to work in the field of entertainment lighting. The industry includes common positions tailored to our industry such as marketing/press, accounting, warehouse employees, strategists, consultants, and more. Each serves a particular purpose with a focus on lighting. Throughout your career, you can easily move between many different positions if you wish.

2 Lighting Genres

The entertainment lighting industry has many facets within which you can find a home in one, or work through many. Most in our industry will find that they work across several genres, but specialize in one or two. For instance, a concert lighting designer may engage in some corporate and festival work but never work in television or theater. There is great crossover of skill sets and procedures between the various genres, but the nuances of each are important to understand.

Theater

Many people start in entertainment lighting through theater. It is, after all, where all this started, and the very root of our industry. Theatrical performances occur in various venues and include plays, musicals, readings, and performance art. Generally speaking, the theater is often the lowest paying genre for lighting professionals. The echelon of theater is of course Broadway and the West End, and some lighting focused practitioners can make a good living working in theater in these environments.

Typically, theatrical production is a bit slower paced than others, while also being highly detail oriented. Theater performances are usually temporary, and the magic of the craft can be an allure for many. Theater can be held within many different types of venues and theatrical productions can also tour.

Corporate/Industrial

Many businesses present staged events such as product launches, press releases, sales meetings, shareholder events, and more. These can range from small events in a ballroom to stadium-sized productions. The genre also includes tradeshow booths, video presentations, showrooms, and other business-focused production environments.

Entertainment lighting is usually a large part of these events requiring workers in all positions imaginable. Generally corporate events have larger funds available regarding budget and salary. Many will also generously pay overtime during long production hours. In addition, the actual shows cross bounds between theater, concert, television, and more. Corporate events are a wonderful learning environment, and can be a single genre for one's entire career.

Touring

One of the great standards in our lighting industry is that of touring. Touring commonly refers to concert touring, but there are also many other touring-style events. By definition, a touring production is designed and constructed to move from one location to another, while presenting a similar performance in multiple locations. The logistics of a lighting rig that must be assembled, utilized, disassembled, traveled, and repeated is unique to this segment of entertainment lighting. Many procedures are unique due to the very nature of touring.

Touring of musical acts is the most common form, but touring can also encompass theater, corporate, houses of worship, television, and sporting events. Whenever the same production occurs in multiple locations, utilizing much of the same gear and staff, it is considered to be touring.

House of Worship

Some form of entertainment lighting usually augments religious ceremonies, services, events, performances, and rituals. Many of the same job positions and principles used in other genres apply with house of worship productions. However, there is an almighty purpose for the event that moves beyond a standard performance or presentation.

Often the religious purpose and beliefs will drive the lighting and determine what is appropriate. In some cases the lighting is seen as a part of the ministry, helping to further the message of the religion. House of worship lighting shares fundamentals with theater, corporate, touring, television, festivals, as well as other lighting genres.

Festivals

Music festivals have become very popular over the past 50 years. From the classic Woodstock to today's EDM festivals, this type of event often includes multiple stages with various acts performing across several days. Lighting technicians, programmers, and designers must work with different artists, managers, locations, and more while remaining focused on the overall event goals.

Festival lighting rigs differ from touring as they are constructed for a short period of time, yet need to remain flexible for various acts. The rig itself may need to be reconfigured multiple times during the event. Lighting professionals working on festivals must understand the unique stresses applied to working with a temporary lighting rig and multiple acts.

Special Events/Parties

This category is rather broad as many events can be lumped into this category. However, there are special events and parties happening all over the world at this moment, and most of them are utilizing entertainment lighting. Almost every lighting company, from small to big, will take part in these types of events. Whether a wedding, bar mitzvah, record release, fashion show, after party, or gala, these events are typically one-off occurrences.

In some cases, very little pre-production will be applied, and a truck of gear and a crew will appear to make the event happen. In others, the event will be expertly planned and prepared. The event industry can be very fast-paced, with multiple changes and decisions coming in from different directions and players. Lighting professionals need to learn to handle the environment and ensure the best event possible given all the parameters and stakeholders.

Television

When lighting for television, many facets of our industry change due to the nature of working towards the image on the television screen instead of for a local audience. The response of the camera, understanding of on vs. off-camera locations, shooting practices, color temperatures, and an eye for detail are all essential elements anyone working on television must appreciate.

With a total focus on the final product on the screen, many standard lighting procedures do not apply to television. Everyone, including technicians, programmers, and designers, must conform to the requirements of working within the television genre.

Film

Entertainment lighting for film is similar to television in that it is focused entirely on the picture for the screen. However, entertainment lighting within a film is often considered a unique specialty look, and not the basis of the cinematographer's primary lighting of the film.

There are cases where entertainment lighting products and procedures are used heavily on a film set to achieve the overall lighting. Usually these are handled by lighting professionals focused on the film genre. Rarely do these people cross over into other genres, but it is more common for others to come into film from other facets in our industry. For instance, many Broadway lighting designers have provided theatrical lighting to portions of films.

Nightclubs

Nightclub lighting runs the gamut from audio-triggered fixtures to full complex productions. Some venues will employ lighting professionals, while others will simply have a DJ or staff person playback lighting looks. Nightclubs can be a great opportunity for anyone starting out to hone their lighting craft. With larger clubs and venues, a successful career can be had with increases in responsibility and creativity. However, most people tend to move on from club lighting to other genres.

Theme Parks/Cruise Ships

Many facets of our industry fall into a far-reaching range of entertainment with varying forms of production. Two common areas are theme parks and cruise ships. Entertainment lighting professionals will find that these venues have production types that cover the extent of our industry. There are opportunities to work in any position and nearly any of the genres within a single theme park or cruise ship.

Sporting Events

Hockey, basketball, football, Esports, and other sporting events utilize entertainment lighting to thrill the fans and present the players. Often lighting staff are employed to install, program, design, and operate this lighting. It is very common to find entertainment lighting products built into scoreboards, stadiums, dashers, and other locations around the sporting event. While the lighting is secondary to the sporting event, often the production is highlighted as an event in itself.

Architectural Installations

Buildings are being lit inside and out, and often the tasks of entertainment lighting workers are called in to assist. An entirely different industry of architectural designers exists outside of the entertainment lighting field, but they will often call upon our field to help with specific projects.

Working in the architectural genre is different from others as the timelines, players, and purposes are vastly distinctive. Usually design and programming services are required, while technician work is often delegated to installation contractors. Projects can take many years to come to fruition, and long-term maintenance is usually required once the project is running.

Manufacturer Demos

Although not exactly a genre of production, it should be noted that lighting manufacturers often create elaborate light shows and demonstrations. As the focus changes from lighting a production to the lighting fixtures being the selling point, so do the creative and technical challenges. While many of these events are orchestrated within the manufacturer's staff, in some cases they bring in outside designers, programmers, and technicians.

When the production exists to promote lighting fixtures and their features, then the requirements of the lighting design and rig change as well. These demos tend to focus more on the output and abilities of the products and less on the artistic qualities of real-world usage.

3 Lighting Skills

First and foremost, to work in the entertainment lighting field you must have a general understanding about lighting. No matter your experience level or current role, you need to be familiar with why and how light is used in the entertainment field. Even the person in the shop loading a truck with truss and roadcases should have a basic knowledge of how the gear is going to be used within the production.

There are, of course, people pushing roadcases and hanging lights who have no information in their brains about lighting, but generally this is not a good idea. If you plan to make a career within the industry, then it is imperative that you learn all you can along the way. If you are starting out working in a shop, ask questions, read books, watch videos, take classes, and absorb all as much as possible about our business.

Early in your career you will be a generalist, studying various facets of lighting. By learning as many different aspects of lighting as possible, you will be able to home in on which path you wish to take going forward. Then once you are ready to focus in on a particular position, start learning as much as possible to become a specialist in that area.

Once you narrow down your goals, keep moving forward but learn all you can about your desired position. For instance, if you wish to become a lighting designer, start looking at current designs and try to recognize the designer's choices. Don't criticize, but instead question and learn. Practice at every opportunity with actual consoles, lights, and other equipment. Ask questions, and become a sponge regarding lighting design information. Most of all, don't be in a hurry to hit it big. As with any career, large successes come after long periods of experience and learning. You probably won't jump overnight from hanging lights to being the lighting designer of a major production.

Spin a Wrench

Most people in our industry start out as lighting technicians in school or just out of school. During this time, they learn to hang and circuit lighting fixtures, focus, roll cable, push road cases, and much more. This physical labor is key to gaining a working knowledge of entertainment lighting. No matter where your career takes you, these experiences will assist you in your duties and responsibilities later on.

For example, LDs who have hung lights and loaded trucks design rigs that tend to work better on the road than those who have not. The best lighting manufacturer engineers worked in theater or concerts while earning their engineering degrees.

Get Real

There is no substitute for working with real lighting. Many inexperienced designers and programmers will make use of visualizers to perfect their skills and practice various procedures. While this is useful for building muscle memory and mocking up concepts, it can also have very negative consequences. By only working in the virtual world, you will never understand the various constraints that are always present in real life.

With a virtual rig physics, budgets, time, human abilities, power, other departments, and a multitude of additional factors do not limit you. When building up your experience bank, nothing is better than real-world occurrences. Say yes to as many opportunities as possible that will present you with real-world experience. Remember, YouTube videos demonstrating your programming or design abilities on a visualizer are not replacements for actual lighting, and should never be promoted as such.

Lighting Skills

Light is a physical part of our world and you should take every opportunity possible to learn and understand about light as it pertains to your current job and your career path. You cannot expect to run the lighting in a church and then a year later be the lighting designer of a major television show or concert tour. Table 3.1 details essential elements of lighting of which everyone in our industry should possess at least a basic knowledge. Depending upon your position and goals, you may choose to become a specialist in certain aspects, or even as many as possible.

Learning Resources

When you have the drive to learn about lighting or increase your knowledge about lighting, there is a vast number of resources available to you. Starting at a young age, some schools teach lighting skills from junior high forward. Many colleges have programs dedicated to entertainment lighting design. Numerous production and lighting companies have internship programs allowing you to simultaneously learn and earn.

Table 3.1 Lighting Skills to Learn

Fixture types and purpose	Fresnels, ellipsoidals, automated profiles, birdies, and more. Understand and know the differences, similarities, uses, brands, sub-types, and availability of the various fixtures.
Hanging fixtures	The skill of attaching fixtures to pipes and truss, running cables, understanding power and data distribution.
Focusing fixtures	Aiming and adjusting fixtures manually or remotely. Maximizing usage potential and light distribution.
Cabling and electricity	Safety procedures, electron flows, connections, current draw.
Safety procedures	Every job should look out for safety of all concerned. Recognize standard safety procedures regarding designing, hanging, connecting of fixtures, as well as handling emergency situations. Be aware of dangers due to light or lack of light.
Control protocols	DMX, Art-Net, sACN, RDMnet – what are the differences and similarities? When to use each, how each affects signal flow, rig setup, and programming.
Automated lighting programming concepts	Standard procedures and terminology, programming flow, playback procedures. Basic console procedures and syntax.
Design aspects	Angles, shadows, brightness, color, emotion, timing, and creativity.
Cueing	Structure, purpose, labeling, timing, numbering.
Building looks	What defines a look? Creativity vs. function.
Color theory	Additive vs. subtractive, complementary vs. secondary, color rendering, color temperature.
Timing	Understanding musical composition and theory, working with time-based systems such as SMPTE, natural and unnatural durations regarding light.
Drafting and visualization	Computer programs, standard procedures, communication, concept vs. reality. Working in a 3D realm.
Metrics of light	Measuring and comprehending footcandles, lux, lumens, and other measurements. Understanding color rendering, color temperature, efficacy, and luminous flux.
Physics of light	Particles, spectrum, waves, photons, and radiation.

Plus, there are abundant books, seminars, videos, articles, and other opportunities for you to fill your lighting sponge with information. Table 3.2 explains some of the most common resources for learning and enhancing your lighting skills.

Table 3.2 Lighting Knowledge Resources

Books	Several publishers create subject matter related to entertainment lighting. A sample includes: Focal Press – www.routledge.com/focalpress Silman James Press – www.silmanjamespress.com Entertainment Technology Press – www.etbooks.co.uk
User manuals	Every product comes with instructions. Reading user manuals will fill your brain with valuable information. Mostly found in PDF format on manufacturer websites, also commonly as a help file in software.
Industry press	Much of the industry press provides articles and other resources related to learning about lighting. A few include: PLSN – www.plsn.com Lighting & Sound America – www.lightingandsoundamerica.com Lighting & Sound International – www.lsionline.com Live Design – www.livedesign.com Technologies for Worship – www.tfwm.com
Schools	High schools, colleges, universities, and trade schools around the world provide lighting programs.
Internships	Production companies such as PRG, Upstaging, and Clearwing, as well as manufacturers such as Martin, Robe, and High End Systems offer paid internships.
Manufacturer trainings	Most lighting manufacturers offer training sessions detailing their specific products. Additional training resources are usually available on their websites too.
Tradeshows	Tradeshows are a great place to learn about new products as well as attend learning seminars and workshops. Many different entertainment lighting focused tradeshows are held around the world. Here are a few: LDI – www.ldishow.com USITT – www.usittshow.com ProLight and Sound – pls.messefrankfurt.com PLASA – www.plasashow.com Showlight – www.showlight.org
Online	Numerous professional and amateur websites offer information in the form of blogs, videos, podcasts, and more. Product tutorials and demonstrations are often found on manufacturer websites.
Seminars and workshops	Manufacturers, lighting/production companies, unions, and other institutions host learning seminars or workshops, often for free.
On the job training	Every show is a learning opportunity.

4 Organizational Skills

The entertainment lighting industry has many positions, and the majority of them rely on strong organizational skills. Whether a lighting designer, programmer, technician, or salesperson, one must learn to organize documentation, workflow, equipment, and personal business.

Documentation

Around the world, documentation and data storage are key to nearly every business. Within our lighting field, documentation is used not just for archival purposes, but also to convey information, share resources, and equalize communication processes. Lighting designers and their assistants produce the most documentation in the form of light plots, magic sheets, tracking information, cue sheets, and more.

Lighting programmers tend to keep their documentation embedded within the console show files instead of creating additional paperwork or files. Lighting crew chiefs and technicians produce documents such as hook up and patch sheets, repair procedures, shop orders, truck pack lists, and more. Sales personnel often document opportunities and contacts, while manufacturing staff have a plethora of documentation involved in the design and manufacture of new products.

Right from the Start

With all these various documents, lighting professionals must find an organizational system that works best for themselves. Some prefer to build a binder per show, in which they place hard copies of all documents, plots, and other materials. Others turn to a total digital world and instead utilize folder systems within their computer to organize the various documents.

Whether digital or analog, you must start with yourself to get organized on a per show basis. First, create a folder or binder with a name of the production. Within this parcel, make sub folders or tabs to segment the information by type. Some common examples include: plots, renders, patch, cues, show files, shop orders, scripts/show flows, staff info, procedures, timelines, and travel documents. Fill each segment with the appropriate information and update as required. Place out of date versions of documents in further sub folders labeled with the word "old." Doing so will keep the most current documents in the main folder, but still allow you to access the previous versions if required.

This method will allow you to quickly and easily find any type of document per production. In a similar manner, you should organize your own business documentation. I have a folder on my computer titled "Invoices," which then has subfolders by year. Each year has copies of the invoices I have created and sent to my clients. In a similar manner, I also have folders containing yearly subfolders regarding my corporate business such as tax forms, letters of agreement, receipts, etc.

Email communication moves the world, and it is important that you can sort through your emails to determine what is currently significant and find older information as well. A good plan to utilize folders and maintain your inbox is essential for a structured email system. I personally have email folders designated for production mails (including subfolders by production), personal emails, travel documents, receipts, business, and more. You need to be diligent to move emails into folders as they come

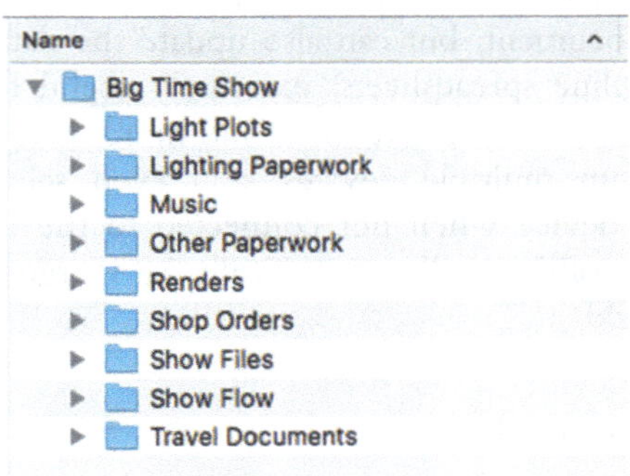

Figure 4.1 Folders for a Production

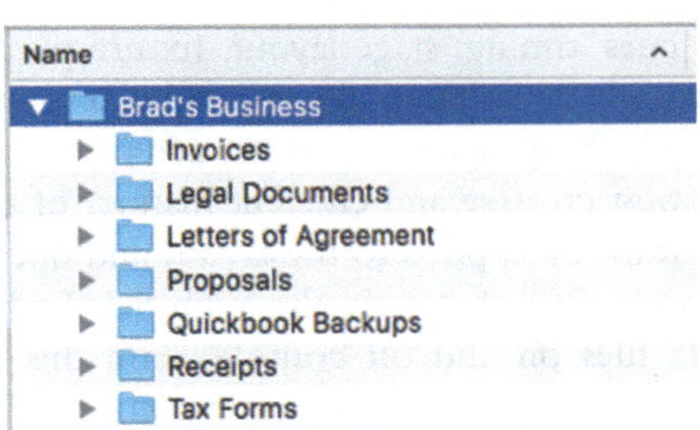

Figure 4.2 Folders for a Business

in, otherwise your inbox will become an overloaded pile of mish-mash that you will come to regret.

Shared Organization

When communicating with others regarding the lighting of a production, organization is key to simplicity and reduced confusion. Many apps and tools exist to assist with organized sharing, but they all require dedication and perseverance from the user. A structured mindset is essential for success with any of these tools. File sharing services such as Dropbox and Google Drive are wonderful resources as they serve as a repository of files that can be shared and synchronized with ease.

When using a file sharing system, be sure to label all folders clearly and use a similar structure to that described above. Don't just lump all the documents for a show in a single folder and hope that others will be able to navigate through the pile. Instead, create subfolders labeled by category and disseminate the data accordingly. Some people will even place a simple text file in the main folder that has instructions detailing the organization plan for the shared files.

Many services such as Google Docs also allow for online shared documentation. This can be extremely helpful, as multiple people not only have access to a document, but can also update the data as needed. I find Google Sheets (online spreadsheets) extremely useful for agendas, itineraries, inventory lists, cue sheets, renders, and more. Plus you can access the documentation from multiple devices, and even select to have copies available on your device when not connected to the Internet. Then any changes you make offline will automatically be merged when you rejoin the connected world.

A Programmer's World

Most lighting programmers are found to have a high level of organizational skills, as the simple act of creating console date requires management. They must create screen views that provide various types of information segmented by purpose while also allowing quick access and simple workflow. Furthermore, data such as cues, timing, stage layout, fixture purpose, and more are all key informational elements that cannot just be thrown haphazardly into the show file. Instead, the programmer must determine the best methodologies to allow the most creative and efficient manner of programming.

Careful thought must be applied to the layout and operation of playbacks on a console and archiving of data is also essential. Lighting programmers routinely store data files on and off consoles, and this should be accomplished in a manner that is well defined and organized. This process usually consists of detailed labeling of show files as well as methodical storage locations and folder systems.

Beyond Data

Organization is not just about data and documentation. A good plan is essential for attacking tasks, and the more thought you put into a plan, the better it will be executed. When you have many duties to achieve a goal or complete a production, first make a list of the tasks entailed. Then work from the top, attacking the next right thing to get closer to your goal.

For instance, if you are a technician or crew chief, learn as much about the production as possible and then craft a plan to get from shop order to load in and beyond. The preparation can be in your mind or written down to share with others. Proper planning is a key element of organization for any series of tasks.

Personal Organization

In addition to data and documentation for productions, you should try to systematize matters in your own life as much as possible. Ideally a personal backpack, roadcase, or workspace should allow you to quickly find what you need, and it should maintain a clean indispensable inventory of necessities. If you have to spend five minutes digging through your bag to find a pen, or open multiple drawers on your workbox to locate a roll of gaff tape, then you need to reorganize.

Think through your daily workflow and needs, and place common objects in easy to find locations and label drawers and pockets as needed. Be sure to return items back to their original location and replenish things that get depleted. Beyond your lighting work, look to determine how you can better organize objects and data within your life. Your home, smart phone, suitcase, and car are all places to arrange and simplify. When you live and work an organized life, everything becomes easier and more efficient.

5 Working Requirements

Depending on your current position or job duty, there may be certain requirements for your employment. Degrees, experience, certification, and references are commonly essential for specific jobs.

Experience

It should go without saying, but your past experience will be a key asset for your future employment. If you have only ever worked as a lighting technician at your local church, then you probably won't get a job tomorrow as a lighting designer for a major concert tour.

When relaying information about your lighting experience, be honest and concise. In many cases, you will have worked on countless different productions and do not need to list every show when applying for a new job. Instead, clearly explain your timeframe for working in your current position. For instance simply state, "Lighting technician for company X from 2017–2020." Then, when asked about specific shows, explain favorite or important productions in person. No one wants to read a long list of shows that they may or may not have heard of.

You should, however, be able to explain your duties and abilities for the various productions, as well as name people you have worked alongside. If working as a designer, you should put together a portfolio showing your work and list recently designed productions. Digital photos, plots, videos, and more will be expected to demonstrate your creative abilities.

References

As our business is highly relationship based, your references will often be very important to your career. When possible, list or suggest references that you worked with in the past. Remember too that in life it is never a good idea to "burn bridges." You should try to always be amenable with all those that you come into contact with. You never know who will have input into your next job or gig. When you work with key industry people, ask them if you can list them as a reference before you put their name on your resume. Remember too that our industry is relatively small, so chances are that the people reading your references may know the names you list. Never make up or exaggerate your relationships, and always give thanks to those that have helped you in your career.

Certifications

Within the entertainment lighting profession, there are a few certifications that can be acquired. Most require extreme study followed by comprehensive testing. Once you earn a certification, be sure to list it on your resume.

ETCP

Entertainment electricians can gain certification through the Entertainment Technician Certification Program (ETCP). The ETCP mission statement clarifies their goals:

> We endeavor to develop a Personnel Certification Program to the highest standards that recognizes individuals who have demonstrated knowledge, skills, and abilities in specific entertainment technology disciplines. By providing a thorough, independent assessment of aptitude, ETCP strives to enhance safety, improve performance, stimulate training, reduce workplace risk, and give due recognition to the professional skills of entertainment technicians.

Some employers will require current certification with ETCP, while others will just view it as a plus on your resume. Either way, the training is a valuable tool for the safety of all, and should be considered by anyone working in the entertainment lighting field. Several approved trainers and courses have been established and testing is held at regular intervals. Full details can be found at etcp.esta.org.

Manufacturer Training

Many manufacturers of lighting equipment will offer training classes teaching attendees how to repair lighting fixtures. In order to comply with warranty status and to teach proper techniques, these seminars often come with a certification status. Once certified, technicians can confirm to employers that they have been trained in the proper repair methodologies specific to that equipment. As every manufacturer is different, multiple certifications may be required depending upon your job.

Manufacturers of lighting consoles and media servers often hold training sessions to teach the programming and operation of their consoles and servers. Rarely do they provide any type of certification other than a "completion of course" document. This is because they only teach the specifics of their product and not how to actually apply the knowledge for a production. You should never assume that because you attended manufacturer training that you also became a certified programmer. However, these courses will provide you with in depth knowledge of the operation of their products.

OSHA

The United States Department of Labor runs the Occupational Safety and Health Administration (OSHA), which oversees workplace safety on many levels. Just as with any industry, OSHA guidelines must be followed in the lighting field as well. Furthermore, many states, unions, and employers require OSHA training and certification by their employees. Within our field this is generally the OSHA 10 and OSHA 30 programs. Learn more about OSHA trainings at www.osha.gov.

These courses cover topics such as personal protective equipment, materials handling, elevated platforms, working surfaces, fall protection, electrical safety, and emergency planning. OSHA procedures for obtaining the required information and completing the testing are very well defined. Many websites offer official courses and testing as well as unions, lighting shops, industry tradeshows, and other organizations.

Recertification

Remember that most certifications are only valid for a specific period of time. You will likely need to recertify every few years to maintain your status. Always be sure to check the terms when applying for certification and then be sure to re-qualify as needed.

Union Requirements

In order to join a union, you may have to pass a test, attend trainings, pay an initiation fee, or demonstrate your skills. Each union and local office will have its own set of prerequisites. Some of these will be unique to the union, while others will involve nationally recognized certification programs. In many cases, the union will cover the cost of industry certification for its members.

Degrees and Experience

Many positions within the entertainment lighting industry will require the employee to have a college degree or equivalent industry experience. In many cases, the degree requirement is not as important as actual practice. Always ask the employer if your working background will fulfill the experience requirement in place of a degree. While a degree can be beneficial to a person's personal development, education, and resume, it often does not lead to additional employment over industry experience.

International Regulations

When working outside of your home country, you will likely need to apply for and obtain a working visa. In many cases, the production will arrange this matter for you. Under every circumstance, be sure that you know the laws when working outside your home country. The right to work, taxation, and other labor requirements could cause substantial headaches if not followed accordingly.

I was once almost not allowed into a country simply because I did not have the proper paperwork. I had to pay a hefty fine at customs and then have the production reimburse me afterwards. However, the trouble continued for another ten years as my file indicated the violation each time I subsequently entered that particular country. Even though the specific show was long over, my misfortunes persisted. Always learn and follow employment regulations when working abroad.

6 Ethics of Lighting

When it comes to working in the entertainment lighting field, the basic principles of conduct and professionalism for any occupation apply. However, there are also several unique circumstances and instances that should always be followed.

You Are ALWAYS Part of a Team

In our industry there are absolutely no positions that operate independently. No matter what your role, you will find that you are always working with others. It is very important that you continuously cooperate with your teammates and keep your ego in check. Furthermore, there are only extremely rare circumstances where the actual lighting *is* the show. In most cases the lighting is just one element in a larger production. Usually lighting's purpose is to provide illumination, while many productions also use lighting to enhance the performance.

Remember the goal of the show and the role of lighting within. Perhaps the artist or producer does not want to spend additional money for another truck of gear, or maybe the director prefers to not see beams of light in haze. While more gear or haze might be important to you, the impact on the show is more important. Decisions such as these must be tempered with an overriding understanding of lighting's contribution to the final result.

Always keep in mind that you are part of a lighting team. At all times recognize your teammates and, no matter your role, never simply claim that *you* "did the lighting" on a specific show. Acknowledge your position, but give credit to your crew, the lighting shop, the director, the artist, and all others involved. Try to use "we" statements when talking about the lighting for a show.

Avoid Title Confusion

The title lighting director is often shortened to just "LD." This of course is a common moniker for the term lighting designer. Confusion can easily occur when a lighting director identifies him or herself as the show's "LD."

It is very important that anyone in a lighting director role clearly refers to him or herself as a lighting director. Artists, management, producers, directors and other production staff may not understand the different lighting positions and titles. Claiming the "LD" designation should only be reserved for the lighting designer.

Furthermore, in the television world a lighting director title often refers to the person in charge of the lighting (aka lighting designer). It also could be used for a lighting programmer or a console operator. When working in television, it is very important to be clear about position responsibilities and not just use the term "lighting director" loosely.

Ask for Help

A growing career involves continuous learning. You should always ask for help when you are unsure how to achieve or accomplish something assigned to you. There is always someone who has encountered the same situation and that is willing to assist or teach you. Seek out guidance and support whenever needed.

Some will tell you to "fake it until you make it," but I find this to be lousy advice. Too often people will take on something that they are not prepared to handle. This can lead to dangerous situations, unfair costs, and incomplete tasks. If you are not sure how to accomplish something, make it known that you need assistance, read a manual, or ask a peer for advice.

Safety First

The very nature of a lighting rig entails positioning thousands of pounds of electrical equipment above and around a stage. Safety for all concerned must constantly be paramount at every moment. Never allow safety to slip in favor of price reductions, time savings, laziness, or other reasons. Safety is everyone's responsibility and must be adhered to above all else, even the show occurring.

Look at your position within the show and see where you can apply safe practices. Leave the rigging tasks to the rigging crew, but do speak up if you have concerns. In addition, think about general safety for other staff, the performers, and the audience. Situations outside your control such as weather, terrorism, and natural disasters can all be mitigated with safety planning and implementation. Keep safety at the forefront of all you do, no matter how minor.

The Event Safety Alliance (ESA) is a non-profit organization committed to promoting "life safety first" in all aspects of event production. Their programs, trainings, education, and resources all center around safety in the entertainment industry. Learn more about safety and the ESA at www.eventsafetyalliance.org.

The Show Must Go On

Since the early 1900s, the phrase, "the show must go on" has been used in the entertainment field to remind all involved that the show itself is the most important factor of any presentation. This axiom should live inside you as a reminder to always do your part to ensure the success of the show. It is amazing to see a production team come together in times of crisis to carry out this mantra.

For instance, I was working a corporate event as a programmer and lighting director with a major band about to take the stage. There was a five-minute video that played before the band's entrance. Just as the video started, the crew chief alerted me to the fact that the building's power had a problem and they had to disconnect the lighting distros and connect them to a different electrical service. As the video played, we doused the rig, moved the electrical connections, fired up the rig, and were ready to go right as the video ended.

During the electrical downtime, I quickly devised a plan as to how to continue the show in the event that there was no power to the lighting rig. I spoke with my followspot operators and prepared them and the stage manager for the possibilities of a band without a lighting rig. Luckily we did not have to revert to this contingency, but we all were dedicated to continuing the show; and in fact the show did go on as planned.

Remember though, that "safety first" is the most important truth and thus should always trump "the show must go on" adage. Never allow a production to continue with blatant dangers to life and limb.

It's Someone's Big Moment

In our industry we get to work on many different types of productions, and they each have their own stakeholders. The producers, performers, and even the audience each have a vested interest in the production being the best thing they have ever been a part of.

If you are working on a small corporate event with a rig of less than twenty moving lights for a local dry-cleaner company, the show may seem boring and redundant. This should not deter you from giving it your all and making it the best show possible. Remember, the people on the stage and the people in the audience *want* to be there. Use your craft to make it the magical moment that they are expecting. Honoring that the production is "someone's big moment" will provide purpose and focus to your own lighting work.

Respect Intellectual Property

Designs, concepts, ideas, music, videos, software, and private information are each property of their creator unless otherwise released. Through the course of your job you will likely be exposed to many things that you should not freely share or use without proper permission.

Common intellectual property that you will encounter includes released and un-released music, custom and stock video content, as well as scripts or product details. It is your duty to only utilize these items as allowed through the course of the production. When you have been entrusted with content, you must adhere accordingly or face severe criminal and career penalties. In some cases, you may even be asked to sign a legal Non-Disclosure Agreement (NDA). This document will bind your secrecy regarding a project's details. Always respect the rules set forth within the agreement.

Intellectual property can also pertain to certain lighting products. As in most industries, counterfeit products occasionally are available. In some markets, it is common to find knock-off lighting consoles that are not from the original manufacturer. They may or may not operate using the official software. Additionally, there have been copies of lighting fixtures, media servers, and other lighting products. When presented with an opportunity to use any of these illegal products, you should absolutely refuse them. Only with diligence can we eliminate these counterfeits from our field.

Always be aware of intellectual property and behave in an ethical manner. If you are unclear as to the rights of content, then be sure to ask before distributing. Always give credit where credit is due and follow laws pertaining to patents and copyrights.

Dress the Part

When working on any production in a technical position, stage blacks (all black clothing) are commonly required or at least expected. During load-in/out, rehearsals, and other non-performance times the dress code may be a bit relaxed. If working a front of house position such as a programmer, operator, or designer, then depending on the type of show more presentable clothing might be in order. When running the lighting for an outdoor concert you can usually wear shorts and a t-shirt, but when sitting a console during a corporate event or awards broadcast you should dress much nicer.

I know a young lighting technician who was hired for a high roller's New Year's Eve party at a casino in Las Vegas. He had just come off the road from concert touring and was did not fully consider the nature of the event he was working. When he arrived for work on the day of the show, he was wearing jean shorts and a t-shirt, the same as he had during load-in and rehearsals. Knowing his job during the show was to run a followspot in the ceiling, he did not consider wearing show blacks.

As you probably guessed, he was not well received when he arrived that afternoon for work. He was promptly told that he needed to wear show blacks. He had to go to a nearby store and purchase black clothing at a premium price. When working any show, you should always plan to dress in show blacks, as you may have to enter the stage or the audience areas unexpectedly. Furthermore, you can never go wrong with show blacks at any production, rehearsal, or other event situation.

I also remember a time when a FOH lighting technician came out to the consoles during the pre-production period of a large concert tour I was programming. It was a hot summer afternoon in a covered stadium, and he appeared without wearing a shirt and shoes. The lighting designer asked him why he was walking around like this, and his answer was simply that he was hot. As you can imagine, this did not go over very well, and he was promptly fired from the tour. Always dress the part for the show and your position. Plus consider your personal safety, as being barefoot in a venue is never a good idea.

We All Know Each Other

Our industry is made up of a very small, eclectic group of people. Word gets around rather quickly and can help or hinder your career. If you work hard and are a kind person, then good things will be said of you and your working habits. Remember that burning bridges, lying, cheating, stealing, and other misgivings will continue with you for your career, so don't enter into these negative practices.

Also consider that falsifying or exaggerating your experience is very easy to verify. I once read a resume of a guy who claimed he was the lighting designer for the band U2. Most people in our industry know that Willie Williams has been the band's LD from 1982 onward. The name on this resume was certainly not Willie's and thus I immediately disregarded this person as a potential candidate.

Even if you work with a band for a performance at your venue, festival, or corporate event, this doesn't allow you to claim to have designed lighting for the band. That honor only goes to those who are hired directly for that specific position. Always be clear and honest regarding your career history and experience.

Word travels fast and you want to have positive comments made about you. Honesty, politeness, and hard work are all attributes that we wish to be associated with our personality. By living with integrity, you will be spoken of well.

Sex, Drugs, and Rock and Roll

Most industries have their share of problems due to partying and abuses; however, the entertainment industry certainly is better known for this type of debauchery in many respects. Alcohol, drugs, sex on tour buses, and favors for backstage passes are often associated with the climate we work in. In some circles, the stereotype "roadie" image still rings true, but it is imperative for you to focus on your professionalism. What you do on your own time is your own business (as long as it complies with local laws), but do not ever think it is acceptable to work impaired or abuse your position.

Having stated the strict message above, I can also confirm that in many cases this is not reality. There are those who will drink or use drugs while working on productions. And while most of the time nothing bad will happen, there could be severe consequences. If you are climbing truss, working with electricity, operating motors, or doing anything else that could affect your life or the life of others, then sobriety during work is vital. If you are sitting at a console pressing buttons or a salesperson hosting a dinner, some inebriation may be acceptable in certain circumstances. Use your best judgment here, and always err on the side of caution.

Should you fall into trouble with drugs or alcohol, seek assistance, and do not worry that your newly found sobriety will cause trouble within the field. Sobriety is well respected and has been achieved by many in the entertainment lighting field. You will find compassion and assistance through many. Some tours even offer a "sober bus" where alcohol is not permitted.

When it comes to drugs and alcohol, always use your best judgment and adhere to local laws. Never trade for backstage access, and continually work as a professional. Evaluate the consequences of temptations and make the best choice possible. Remember that people are constantly watching you, and that nothing in our industry is secret.

All Access is Not All Access

Working on a production will give you access to an exciting part of society, the entertainment world. Often you will have abilities to enter areas or see things that the general public only wishes to experience. However, just because you are issued a badge that says "all access" does not mean that you can go into the band's dressing room anytime you want. That badge simply allows you into the areas that you need to enter in order to attend to your job duties.

Furthermore, taking photos, videos, or sharing information about a production on social media or websites is often frowned upon. It might be exciting to be backstage standing next to a famous actor or musician, but this does not give you the right to take a photo of the moment or ask for an autograph. Your job is still your job; you are not there as a fan. There is an unspoken rule that those who are working on a show shall respect the celebrity and any secrecy involved with the production.

Many corporate events may be announcing a new product or strategy. You might learn the sensitive details during pre-production or rehearsal, but must comply with the wishes to remain silent until after the show airs. In many cases, stickers are placed on camera phones or recording devices are not even allowed in the venue. A true lighting professional will obey the privacy needs of the production, without exception.

Sometimes opportunities will arise for you to ask for an autograph or have a special personal moment with celebrities. Even in these circumstances you should remain professional and treat the famous people the same as anyone else working on the production. Often when I am asked what it is like to meet a famous person, I just explain it is no different than working for a company and having a meeting with the management team. They are just the people at the top of the org chart. In some rare occasions, lighting professionals (usually directors or designers) will become very friendly with celebrities. I have always found these relationships to be very personal and rarely discussed outside of very tight circles of friends and family.

7 Best Show Ever

Often, when asked "What is your best show ever?" people in the entertainment industry will answer either "The last one" or "The next one." A good working mantra is to always treat the current production as the very best one with which you will ever be involved. This mindset puts forth an attitude and drive that forces you to give your all to the tasks at hand.

No matter your position or the scale of the production, you should always strive to do your best to ensure the show comes across flawlessly and easily. By giving it your all, you are saying that this show is the most important thing in your career at this very moment. Sure, you might need to complete a plot for your next show, or start thinking about the patch for a future event, but while on this particular production you should remain focused.

To be elite takes dedication and discipline. You cannot just decide that you are the best lighting technician and then it becomes true. You have to dedicate yourself to your craft and be disciplined at working at your best in every situation. Ask yourself often if you are working at a level that would be viewed by others as average, above average, or beyond.

If your job is to program the lighting, do your finest, and offer up some amazing looks that have never been seen before. If you are a lighting technician in the shop, take your time to double check everything and clearly label roadcases, fixture numbers, cable runs, et cetera. As a followspot operator, meticulously check all operations of your spot and headset, and then listen closely to the instructions of the lighting director. The drive to make the current show the greatest ever can reside in everyone involved, no matter your position. This infectious spirit is what drives success on an incredible level and can be found from the smallest to the largest productions worldwide.

The work may be difficult or involve long hours, but with an attitude that this is your best show ever you know that you will get through the tough times. The more staff who are dedicated to this mindset, the better the actual working environment. Most stress on a production comes from interpersonal relationships among those involved. If you are striving to make this show the best possible, then you can easily resolve differences with others that are motivated to achieve the same goal. In some cases, you may need to surrender your opinion for the sake of the production. This will be easiest if you are focused on making the show the most excellent one ever. When you allow your ego to overrule the show's outcome, then no one wins. Remembering that you are part of a team working on this production will help you with your decisions and relations with other members of the production staff.

Doing your best does not mean filling in for other people's deficiencies or always going above and beyond what is expected. Simply giving your duties the utmost of your capabilities is what is asked as a bare minimum. Often helping others or going beyond your capabilities will result in lackluster performance of your primary job. You must be careful to not slack on your main goals in the name of assisting.

For example, imagine you are assigned to program the lighting, but you go on stage to help troubleshoot a faulty fixture. During this time, no programming is getting accomplished, due to your absence from the console. You are no longer doing your best (in your assigned position) for the production. Sure, you are helping the show by sharing your knowledge to restore the broken light, but in the same course of time many cues could have been written into the console. Now the show's programming will suffer, when a technician who is allocated to repair the fixture could have accomplished the task instead. Always use your best judgment to determine how your actions will benefit or impair the production.

Many people in our industry often find themselves working various positions and on different sized productions. You may have just been the lighting director on a major stadium concert tour and then a week later find yourself as the crew chief for a wedding reception. The stadium tour played to 50,000 fans a night with a rig of hundreds of lights, while the wedding reception will be seen only by the 100 guests with a modest rig of 50 fixtures. You should have the same determination for either show to ensure that it is absolutely the best it can be for the audience. Do not come to the wedding reception gig with a big head and reducing its importance. It is equally (if not more) significant to the families in attendance than to the band on stage at the concert tour. Making every production your best show ever will ensure that you are giving your best to achieve your position's goals and contributing to the show's success.

The best show ever is NOT an excuse to skimp on safety. Your best means that everyone involved on the production and the audience is able to leave the event in the same way they came in. Just because you want to do your best and have the greatest show ever does not mean that you climb up a truss in a hurry without clipping in to safety gear. The best show ever should always be equal to the safest show ever.

Those who always put the production ahead of their own goals and needs will certainly have more successful careers. Others on the production will see and feel your dedication to the accomplishment of the show and want to have you participate in future productions. This attitude shines above others who appear jaded or uninterested in the current show. It will be very apparent who is working on the show simply to get a paycheck, compared to those who want the best show ever. When you make every show the best ever you will have a wonderful, prosperous, and fun career in the entertainment lighting field.

Part II

The Career

A career in entertainment lighting becomes more than just working various jobs for many years. Most people have a passion for the business and find that the work is a labor of love, and their career is simply part of their larger lifestyle. Many will say that lighting is in their blood, and that they could not live without it. We are often amazed that people will pay us for something so dear to our hearts.

But simply working within the lighting field requires much more than job duties. A full career requires a greater understanding of business procedures, self-promotion, continual learning, and giving back.

8 A Growing Career

A career is generally described as an occupation undertaken for a major portion of a person's life with opportunities for progress. No matter your career aspirations within the entertainment lighting field, it is important to remember that growth requires time and experience. No one should expect to jump right into major productions without first "paying his or her dues" with smaller shows and lower positions.

A typical path for a lighting professional usually starts as a technician in a lighting shop and then progresses into working on-site with productions. In most cases, one can move from shop, to show technician, to programmer, to assistant designer, to lighting designer. However, this does not suggest than any one position is any less important than another. Along your way, you should always do your best in any position you find yourself, but also look for opportunities to gain experience to help you progress towards your career goals.

Continual Growth

A common method of advancing is to work with smaller productions at a higher position while also maintaining your primary (lower) position on a different show. For instance, you might be a technician on tour, building the rig and fixing fixtures every day. If you have dreams of programming and designing, then you should offer to program and operate the lighting for the opening act. In many cases, this position will be available and may even offer some additional monetary compensation. The additional pay aside, your main goal will be in gaining experience as a programmer and designer. Be careful not to let this additional task distract you from your primary role as a technician, but do your best to gain all the experience possible.

Another method to continually grow is to take all the opportunities that come your way. Many lighting professionals will design lighting for very small productions with extremely limited budgets and even less pay. In order to survive, these professionals also spend much of their time as a technician and/or programmer on various events.

For example, lighting designer Luther Frank has aspirations of lighting shows on Broadway. He has a master's degree in lighting design and moved to New York City right after college. He designs as many small productions as he can, but these do little to help him pay his bills. So he also works as many events as possible in town including corporate events, galas, awards shows, live television, weddings, and more. For these events, he started off working as a technician and quickly moved up to programming. In many cases he also designs these productions in addition to programming.

He averages over one hundred shows a year just working these events. While it is not contributing directly to his Broadway career goal, he is gaining valuable experience and allowing himself opportunities to design lower paying productions in New York City. His roster of contacts from the events world has also led to further projects and shows. Additionally, he finds opportunities to visit, assist, observe, and occasionally work with Broadway lighting designers. He is focused on his design career, while paying his dues and his rent! In fact, with all this event work, he was also able to pay off his student debt in under five years.

Feeding your passion might not feed your wallet, but there are plenty of other positions in the field that will support your financial needs while you work towards your larger career goals.

A Natural Progression

As you start your career, typically you will be a generalist with many different skills. For most, this means working in a shop or on productions as a technician or stagehand. One day you might be loading trucks, and spend another day hanging lights. At other times you might help bolt truss or even set up a console. Your skill set will be based on general knowledge of many different aspects of lighting productions. During this time, you will want to take as many gigs as you can to gain experience and begin to find an area that interests you for specialization. During this phase of your career, it is advisable to say "yes" to every opportunity that comes your way.

As you move along in your career, you should begin to specialize in a specific task or position that is of interest to you. You could decide to become a crew chief, programmer, designer, salesperson, or any other position you desire. While working as a generalist, you should start to focus in on your specialty and learn, train, and practice the required skills as much as possible. You can then selectively choose work that benefits your specialization.

Luther Frank started working any gigs he could get in NYC, then after a few years focused in on programming and design work only. Now he turns down general crew work in favor only of opportunities that allow him to program and/or design. As you progress in your career and decide on a focused path, you need to say "no" to gigs that do not suit your specialization (unless, of course, you simply need the money).

When working as a specialist instead of a generalist, it is imperative that you let all your clients know your role and skills that you can bring to their production. It is up to you to teach them how to see you. If your goal is to be a designer or programmer, then when you are ready, you should only take work that supports this position. You don't want to take a gig as a technician for a week only to discover you missed out on a programming opportunity at the same time. After years of grabbing every opportunity, it can be daunting to say "no" to work. It will feel odd to turn down a week's work or to pass on a tour. In time, however, you will be much happier working in your area of specialization following your career goals.

Setting Goals

A great career tactic is to set goals for yourself many years apart. You might wish to be a technician at a shop for two years, followed by a production technician for five years, and then a venue role followed by a design focus. You are allowed to define your own timeline, but you should be realistic and flexible. There is no way to judge how much time working as a technician prepares you for your next big career leap, but in general you should never be in a rush to reach the top. If you are happy designing and operating the lighting at the church where you currently work, then you might want to continue that position longer than planned until you are ready for another challenge.

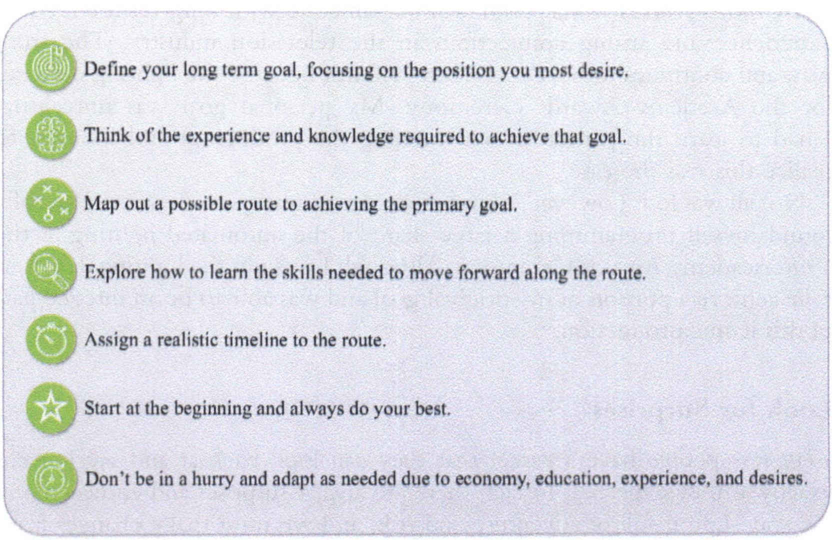

Figure 8.1 Tips for Setting Career Goals

For instance, I had a plan when I first started with High End Systems to work there for five years before returning to the freelance world and focusing on programming and design. At the five-year mark I decided to hold out a bit longer due to new products that were coming to market (consoles and media servers). I felt that it would be best for my career to stay for another year to fully assist in the newest technologies. Then on September 11, 2001 everything changed further when the economy drastically changed in the wake of the terrorist attacks. I decided to continue with the steady employment and not transition to a freelance career for one more year.

Goals are great for creating a plan, but you must be willing to extend deadlines, or make radical changes due to unforeseen forces.

A Real Example

Early in my career, I would study the lighting on the Academy Awards Ceremony with great interest. The production was filled with fantastic automated lighting cueing and was quite a large spectacle. I would record the show and watch segments over and over. Furthermore, I studied the credits to see whom the lighting professionals were that made the show happen. I had dreams of becoming the lighting designer for this production and assigned myself a goal to design the lighting for the Oscars in the next twenty years.

Now, at this time, I did not have much television lighting experience, nor did I understand all the politics and players involved in that particular production. I simply set the goal and thought that I could work hard and achieve it. The reality I eventually learned is that lighting designer Bob Dickinson has lit the show for over 30 years and will probably continue for many more years. His successor will be someone with long-term television experience and strong connections in the television industry. The truth was, and continues to be, that I likely will not become the lighting designer for the Academy Awards Ceremony. My personal goal was unrealistic; I had to learn the particulars surrounding the production and industry to realize this was the case.

Not all was lost, however! Through a series of unexpected events, I actually found myself programming a large share of the automated lighting at the 1996 Academy Awards Ceremony. Although I was not the lighting designer, I *did* achieve a portion of my original goal and was able to be an integral part of this iconic production.

Look for Surprises

Very few people have a career that they can look back at and say it went exactly as they expected. In fact, there are always surprises and choices along the way. Life is full of adventure and risk, and we must make changes from time to time within our career. Some of these changes will be successes and others will lead to failures; but all failures are lessons for growth.

When I joined High End Systems as a lighting programmer in 1996, I had a goal of working for them for only five years and then returning to the freelance programmer and designer realm. However, technology developments, world events, and my personal life caused me to stay employed by them until 2003. With my freelance design and programming plan working well, I received a surprise phone call. Richard Belliveau of High End Systems called to offer me a position within the Engineering team to help develop new lighting consoles and fixtures.

While I had never considered this as a part of my career plan, I did have a bit of experience in the product development field from my previous tenure at High End. I was intrigued by the possibility of contributing to the development of world-class products used by the entire industry. I was torn, though, as my freelance programmer and designer role was going extremely well with lots of growth and opportunities. Ultimately, I took the job, returning to High End Systems with an agreement that allowed me to continue to do a bit of freelancing as well. Over the following years, I learned that I had a talent within product management and this opened new and unexpected doorways within my career.

The entertainment lighting field is full of opportunities, and you should always consider unexpected possibilities. Transitions to companies, joining design firms, manufacturer employment, working with only one band, taking a house gig, starting a rental company, becoming a consultant, teaching others, and developing new products are just some of the changes that lighting professionals have made to their career paths. The choices are up to you, and no matter what you decide you will have a great career within the entertainment lighting field.

Prepare for the Unexpected

Entertainment has been around for centuries and fortunately people will always want to be entertained. However, there is no guarantee that your position or work opportunities will always be available to you. You must prepare for the unexpected by saving money, learning diversified skills, and identifying when to pivot.

Shows come and go, tours stop suddenly, corporate work dries up, and work can disappear completely without notice. At any moment, your upcoming gigs could cancel. It is highly advised to have at least six months of your expenses in an emergency fund for slow times. You might have to tap into these resources to survive the down periods. Additionally, you may need to look for work outside the field to keep income coming in. There is no shame in taking any job if it means the difference between paying or not paying your bills.

When entertainment jobs are not available, see where you can use your skills in other industries. For instance, if you can draft, offer your services to machine shops, engineering offices, and other mechanical design companies.

Perhaps your technical skills can be used in construction, warehousing, or even medical fields. Designers may find related work in art studios, museums, and corporate firms.

In 2020 the COVID-19 pandemic opened everyone's eyes to the fragility of our industry. Beyond just lighting, nearly every single segment of entertainment was put on pause for an unprecedented period of time. Many lost jobs and were forced to scramble to find any possible work. Without any jobs for the foreseeable future, those that could turned to their emergency funds. However, many without any such backup plan instantly found themselves in severe trouble. With little opportunity to pivot to a new market, their suffering became unbearable. Preparing for the unexpected should always be a priority. Remember the unimaginable can occur any time without warning.

During the downtime in 2020, many lighting related learning resources became available to the masses. Industry manufacturers, lighting experts, and schools provided untold amounts of online training and workshops. This became an important time for designers, programmers, and technicians to learn new skills and improve existing knowledge. Resourceful people used the unfortunate crisis as a period to come out stronger. Some even prepared for future events by learning skills outside of lighting to enable them to continue working should future pauses occur to the entertainment industry.

We never know what tomorrow will bring and no show or event is ever one hundred percent guaranteed. Take the time to prepare for the unexpected by saving money, improving skills, and understanding your value in various markets. Never take the current moment for granted and always look ahead with caution and optimism. However, rest assured that entertainment is important to the world and has always continued through the tough times.

9 Attitude Is Everything

Our field is heavily based upon individual relationships, and thus your people skills are some of your most important assets. When I surveyed over a thousand people in our industry for advice they wish to pass on, one theme was overwhelmingly apparent. Your personal attitude and interaction with other humans are extremely important. To put it simply: if people like working with you, then you will work with more people. Never behave as a know it all, smart-ass, jerk, or entitled person. Instead, you should be a friendly team member who does your job well. Otherwise, you likely will not get called back to work on future productions.

Sure, there are times when you need to be assertive and demonstrate your expertise about a situation or procedure. Try to present these moments with kindness and understanding as opposed to simply forcing your opinions or knowledge on others. Safety concerns should always be expressed openly, honestly, but with emphasis. Do not hesitate to point out unsafe situations or potentially dangerous errors. Work with all the teams on a production to resolve problems instead of speaking out against others in a negative manner. Being kind should be your standard mode of operation.

Once my wife and I were at a sandwich shop in New York and ordered two vegetable sandwiches. As the guy was making them, my wife and I talked with him about New York and other topics. When he presented us with the sandwiches they were beautifully made with many vegetables and other goodies. My wife commented about how filled up they were, and the guy told us something amazing. He explained that not all sandwiches are made to the same standard. He said, "If you are nice you get a nice sandwich, but if you are a jerk then you get a jerk sandwich." This nugget of wisdom holds true in many aspects of life, but especially regarding your career. When you are a nice person, then good things will happen in your life. If you are a jerk, then you get what you deserve.

The vast majority of successful lighting people are also the nicest. They have learned how to work well with others, keep their discomfort and stress to themselves, and navigate production politics. They know their skill level and ask for assistance when needed. They are honest with others and want the best for all involved.

I have also met others that have a bad attitude who seem to be stuck in their current role, or they disappear completely from the industry after a while. I was once working a show in a large theater and had a clash with a stagehand nicknamed "Grumpy." He fit his moniker and acted as if he knew more than anyone and did not want to be at work. His negative attitude towards everything and everyone was a detriment to the entire production. Many on the show were having difficulties due to Grumpy's bad attitude. By the third day, the union steward transferred him to another show. Apparently, not long after that, he was told to change his attitude or find a new profession.

Some say that success in our business is based upon 40 percent skill and 60 percent personality. I tend to agree with this formula as I have witnessed it throughout my journey. Treat and respect others as you would like to be treated. Make each experience one that encourages people to want to work with you on future productions. With a great attitude and hard work, you will certainly be asked back to future shows.

Focus on Now

Your outlook will also be better if you are focused on the production and job at hand. Always stay focused on where you are at the moment and try to avoid distractions such as social media, texting, and working on other productions. When your mind wanders from the current show, your stress level will escalate, and your attitude will deteriorate.

Remember that you are paid to be on-site at this production to help make it the best show ever. In return, you should give it your full attention. Any less and you are putting your personal needs and choices above the production. Distractions may alter your mind-set and place you in a negative mood. This will then carry over to your feelings about the show and the people you are working alongside. Instead, save outside influences for non-working hours and keep a smile on your face.

Sure, the rehearsals may be long, the LD might ask for more changes, or the band will change their mind again. These are all stressors that can occur at any moment of a show. As they come to you, try not to say "no," and instead look for opportunities to share in the situation at hand. If you can take changes in stride with a focus on the end result, then your relaxed attitude will flow through to the rest of the team. Plus your "can do" attitude will be admired and respected by those making the requests.

In some cases, time will be of the essence and work will have to happen right away. For instance, if a fixture fails ten minutes before doors, the crew will have to come together to solve the problem and update the new fixture's focus. Even if you had other plans in the moment, this emergency will require your full attention and a good attitude towards solving the problem. Although it might have conflicted with your dinner break or personal time, remembering that the show is the reason you are there will keep your attitude in check.

Thank You

Kindness and gratitude are main elements of a good attitude. If you can be kind to others and grateful for the opportunities presented to you, then you will lead a successful career in the entertainment lighting field. Take the time to thank others daily and always follow up after a show.

I always make it a priority at the end of a show to thank each of the lighting crew as well as other production staff for all their assistance. Furthermore, I will follow up with the people that hired me via a phone call or email, letting them know how much I enjoyed working with them. These simple touches go a long way towards ensuring that people will want to work with you again (assuming you did a great job of course).

We are all very fortunate to get paid to work with lighting in the entertainment business. Even with the various stressors, try to bring yourself back to a state of gratitude for the opportunities to work in this amazing field. Many people spend much of their lives working purely for a paycheck, yet most of us are very fortunate to get paid for a job that we love to do.

10 Training and Learning

Throughout your career you should be open to learning new information regarding lighting. From technology to techniques, there is always something to learn. As your path changes you will need to study your new direction while also improving your personal skills. The industry provides many avenues of education, and each has its own benefits. Generally, a combination of all of the following is best for anyone working within the entertainment lighting profession.

High School

Many begin their entertainment lighting education while in high school. During this time the basics of hanging lights, programming, design and other lighting essentials are usually acquired. Some schools have basic systems with older technologies, while a few schools possess high budgets and the latest of equipment.

The real key in any of these situations is the instructor. Unfortunately, it is common for the technical aspects of high school productions to take a back seat to the acting. In many cases the theater teachers are trained and focused only on acting with very little lighting or technical background. This can lead to students learning on their own or from older students and parents. Look for schools that have a strong technical theater program and experienced instructors.

Colleges and Universities

College and university programs are wonderful environments to learn about lighting. Many schools provide in-depth lighting design and production tracks culminating in BA and MA degrees. As you learn about lighting, you will also receive a basic liberal arts education through a general learning curriculum. Your focus on lighting will provide you with the fundamentals and basic skills to move forward with a lighting career. A good school will provide many diverse production opportunities throughout the school year.

In most cases, colleges will offer a degree program focused entirely on lighting design. While in school you will learn lighting history, design concepts, drafting, practical skills, and more. Because the focus is often on design, you may graduate lacking some of the basic lighting technician or programmer proficiencies required to enter the working world. Upon graduation you might find more employment prospects in these positions than lighting design. Chances are low that you will become a busy lighting designer right out of school.

Countless schools are very theater-centric and may not fully relate to the plethora of lighting opportunities that you will find upon graduation. While theatrical methods certainly are a foundation of most entertainment lighting productions, actual "theater only" careers are rare. Some schools such as City Tech in New York City and North Carolina School of the Arts have stepped away from by offering degrees in production or lighting technology. They utilize the latest products and also include lighting courses that cover various genres of production such as corporate events, concerts, theme parks, and events.

Remember too that in educational environments the students generally learn using only the in-house equipment. This means you might miss out on the latest (or even current) technology being deployed in the field. Look for a program that is not isolated in terms of the available technology. In addition, try to select a school with a large of focus on relationships with outside production staff, creative uses of lighting fixtures, hands-on opportunities, and budgets for rentals. Some schools will even invite in well-known lighting designers to teach classes, evaluate portfolios, or light shows.

College programs are wonderful and can provide a very robust set of courses that are sure to teach anyone the fundamentals of the entertainment lighting industry. A college degree also provides many more life enriching skills that can be used beyond lighting. However, this type of training can take several years and usually costs large sums of money. Completing a degree program could leave you with a large debt to repay.

When looking into a college, it is imperative that you enquire about their equipment use (in-house and rental), number of shows, visiting professionals, curriculum scope, and availability of gear for the students' usage. Ask what production positions are available to you while in school, and be clear about your post-graduation goals. Take the time to find a school with a program that best fits your desire for your future.

Technical School

Beyond traditional colleges and universities, there are several industry-focused schools with degree and certificate programs that specialize in various aspects of our field. For instance, Full Sail University (www.fullsail.edu) in Florida offers a degree in Show Production that includes courses in automated lighting, production technology, and more. They even make available complete educational learning packages that provide students with a laptop loaded with important software as well as other professional tools.

Similarly, the Stagecraft Institute of Las Vegas (www.stagecraftinstitute.com) employs working professionals to teach various lighting courses. They award a seal of qualification upon completion of coursework and testing. Many other technical schools can be found throughout the United States and in other parts of the world as well. Some might find technical schools expensive or limited in their training resources, while others find them a quicker path to employment than traditional colleges.

Internships

A great method to learn and hone your skills is to work as an intern with a production company, design firm, or manufacturer. Most of these positions are paid, so that you can earn money as you learn. Production Resource Group (PRG) has one of the most recognized internship programs in our industry. At most of their locations, they offer a well-organized curriculum that takes attendees through multiple years of employment. As the intern works through various positions, he or she works with a mentor to help guide future career choices. Upon completion of the internship, a position in the shop, on the road, or in production crews is usually offered.

Many other companies such as Upstaging, Clearwing Productions, ETC, and High End Systems have been instrumental in providing internship opportunities to many when they were starting out. The relationships formed while working indirectly for a company are usually life long, and many interns gain employment from their sponsoring company.

On-the-Job Training

Similar to an internship, many regular jobs offer on-the-job training. Training ranges from company required courses to continued education paid for by the employer. Often senior level employees will share knowledge and experience with junior employees. Many venue or lighting or production company positions offer learning opportunities.

Manufacturer Training

Most lighting manufacturers provide training sessions specific to their products. Some of these courses are required for warranty level repairs, while others simply teach how to operate specific pieces of equipment. Often manufacturer training is provided at little or no cost to end users and can be very valuable in understanding a product or range of products.

For instance, lighting console manufacturers regularly teach courses about how to operate their desks. While these courses do not teach the procedures and skills of programming (see my book *The Automated Lighting Programmer's Handbook* for that information), they do teach the required keystrokes, setup, and operation of their consoles. The course is relevant to most anyone working in the entertainment lighting field, as it will provide a basis of understanding about that specific console line. A technician may be called upon to use a desk to test products, while a designer might want to understand the capabilities provided to his or her programmer.

Manufacturers commonly provide training classes for product repair or installation. Many of these classes are extremely technical and include testing and certification. Some will even have pre-qualifications to ensure only the correct people attend the sessions. These trainings are usually very beneficial as they provide insight and skills into products that often are not available from any other source.

Manufacturer training occurs at many different locations. Often the courses happen at the manufacturer's offices, but also can be found at tradeshows, open houses, union sponsored events, and more. Always check the manufacturer's website to see what courses are available and in what locations.

Learning Sessions

Beyond manufacturer-based trainings, there are many opportunities for learning within our industry. Learning sessions, forums, panel discussions, and case studies are common parts of every lighting tradeshow. Since 1988, the Live Design International (LDI) tradeshow has included over a hundred training sessions annually. Many of these are available at little or no charge, while others come with a hefty price tag. Additional tradeshows such as USITT, PLASA, and ProLight & Sound have similar training opportunities.

Learning sessions are also found at production company "open house" events, manufacturer demos, houses of worship, union halls, and even in venues. Often an institution will bring in working professionals to share their knowledge and skills with attendees. These opportunities provide valuable resources and information for students and professionals alike.

User Manuals

A great place to learn more about a specific piece of equipment is to read the user manual. It may seem simple, but most people skip reading the boring guide that comes with a product. However, I can assure you that if you take the time to read through the manual for a new fixture or console that you will find some new information that proves valuable for your next production.

Lighting equipment is very complex, and each piece comes with its own software, hardware, and unique abilities. The user manual will describe all these elements in detail and may even alert you to dangers that you did not know existed. It is always a good idea to read through a manual for any new product that you encounter. This is one of the best ways to get the most from the products you are using. Why would you want to miss out on important features or abilities just because you did not take the time to read and learn about them?

All user manuals are available from manufacturers on their websites. You can download the documents and read them on your smart phone, tablet, and/or computer. Keep the files organized on your platform. Remember to check for updates as products and their manuals get revised over time with different software features and hardware changes. Always download and read the current manual and DMX chart for a product you have never used. This practice will ensure that you are aware of features and procedures related to that specific equipment.

Peer Groups

Our industry is filled with people at all different stages of their career. Everyone has something to share with others. Take the time to sit down with colleagues working at the same level as you and discuss your skills and knowledge. You will likely learn something new while sharing your own experiences. This type of knowledge exchange is always best in person, but can also occur via social media, blogs, email, or even at industry events.

Don't wait for these opportunities to come to you. Instead, take charge and invite others to meet for coffee, and ask questions at a gig when applicable. For instance, if you have a desire to become a programmer, ask the programmer on your next gig if you can watch him or her work. Once you have completed your assigned tasks, sit quietly at FOH and observe the process. Most people are willing to share their knowledge and will be happy to have you shadow them while they work. You can also arrange meet-ups away from productions to further your conversations with peers.

Private Training

There are many people with vast skills and experience in our industry. You can easily reach out to any and ask for training and education. Some may train you for no cost, while others will ask for payment. Either way, you will likely gain valuable information that may not have been available to you any other way.

For instance, long ago I met a technician named Heath Marrinan while I was programming a tour. He was fairly new to programming and saw that I was very experienced. He reached out and inquired about hiring me to personally train him. He invested in his future by flying me to his location, arranging gear, and paying my rate to teach him for three days. He was then able to take this knowledge and expand with practice to become a very accomplished programmer and designer. The private training helped to accelerate his knowledge, and also fostered a great friendship that continues to this day.

Some organizations such as unions, rental companies, houses of worship, or schools may hire professionals to teach their staff important abilities. Many freelancers offer their training services, and a few have even created companies focused on training. For instance, Richard Cadena specializes in various types of lighting training with his Academy of Production Technology (www.aptxl.com). He has traveled the world teaching diverse courses to anyone that chooses to hire him.

Media

Educational materials can be found through our industry media as well. Publications such as Lighting & Sound America, PLSN, and Live Design create content that ranges from magazines to webinars. They employ industry experts to share details about productions, how-to articles, and other learning opportunities. These resources are valuable as they are usually timely and in line with current industry trends and products. In all parts of the world there are similar publications such as Lighting & Sound International, Sound & Light, Show Technology Asia, and TCI.

Beyond the industry press, many online blogs, websites, YouTube videos, podcasts, and social media groups are filled with educational material. Remember, though, that not all of it is curated, and thus it is always important to consider the source. Look for sites that belong to respected industry professionals or others that have already been vetted regarding their experience and knowledge.

Books

Since you are currently reading this book, then hopefully you already understand the importance of the printed word. Many books on the topic of entertainment lighting are available from numerous sources. Take the time to find the publications that cover the topics you need to learn and read the latest editions. Most books are now available in traditional printed format as well as electronic versions. Amazon and other online retailers make it easy to find books about entertainment lighting.

Self-Taught

I have spoken with a few lighting professionals who claim they are "self-taught." However, I highly doubt that this is exactly true. While they probably mean that they never attended a school or seminar, they likely had to consult peers or read manuals to gather further knowledge. Remember, that if you really want to exceed in this industry then you are going to have to go beyond yourself and learn from all the above mentioned resources. If you simply limit yourself to what you can figure out by yourself, then you will certainly be missing out on many key bits of information.

Learning opportunities are available throughout our industry. In order to have a very successful career, you must continue to educate yourself. Never think that you are complete in your education; instead look for learning moments to help you advance forward (or simply keep up with technology).

11 Promoting Yourself

The entertainment lighting industry is actually rather small compared to many other industries. Because of this, word of mouth and reputation are key factors to gaining employment no matter the position you desire. While a resume is essential for some jobs, it is merely a formal step in the hiring process.

Throughout your career, you will need to continually promote yourself, your abilities, and your desire to move forward. There are countless avenues upon which this promotion can take place, and it is very important that you make use of as many as possible. You are your own brand, and only you are responsible for promoting yourself. Remember that you must first do the work, and then promote yourself. Don't publicize yourself as a lighting designer if you have never worked as one.

A Simple Resume

When building a resume for any position, it is generally advised to keep it to a single page. This is true within our field as well. Many try to list all the shows they have been a part of, and this just becomes a waste of time and space. Instead of listing every production, highlight a few key shows in an introductory paragraph.

Then be clear about the types of positions you have held and the skills that you possess. This level of detail is far more useful to a potential employer than reading the title of a show or name of a band. If you have worked as a tech for twenty different tours, simply list the years you have been touring.

Education, special acknowledgments, and unique abilities should also be noted. A few references are always helpful too. Remember to keep your resume simple and to the point. It is not this document that will earn you a job, but rather your skills, personality, and references.

Portfolio

Design positions often require a portfolio that can represent your work and creativity. A good portfolio will include images, videos, plots, conceptual drawings, and more. A portfolio might be a printed document, a large display, or a virtual online experience. In some cases, designers will send their portfolio via a USB stick or on a tablet device.

Along with a resume or portfolio, a designer will often include a list many of their most recent designs along with details such as the director, producing company, and dates. If you have been working for many years, there is no need to list all your past shows. Instead, feature the most recent as well as a few favorites or stand-outs from your career.

Website

It never hurts to have a website to promote yourself and your work. It is one simple place that people can find out what you have worked on, see samples of your work, and gain contact with you. There are many companies that allow you to have a free site (with some advertising), or you can choose to pay for hosting and build your own site. Furthermore, it can be helpful to register a custom URL such as www.bradschiller.com to make it easy for people to find you and remember your name.

Your website should have a clean, modern look, detailing what you do and the types of jobs you want to gain in the future. If you are a designer, having many images and videos of your work is essential. If working another aspect of the industry such as a technician, programmer, or crew chief, you might want to include links to YouTube videos of shows which you were involved. In all cases, be very clear as to your role on the show and credit the lighting designer and other important people and companies.

Your website needs to also have a "call to action" focus that requests viewers to contact you, read your blog, or follow your social media. If the site is merely a portfolio of your work, then visitors will be simple viewers. You want to provide them with as many opportunities to learn about you and your work as possible. A blog or video portal that shares your experience and knowledge will entice people to frequently visit your site.

Search Engine Optimization (SEO) services are big businesses in today's Internet world. They will help ensure that your webpage is chosen among the many during an Internet search. Whether you choose to hire these services or not, you need to take advantage of SEO tools regarding your website. Keywords, metadata, long semantic indexing, and localization are all primary SEO principles to incorporate into your website for maximum results.

Also, don't forget to optimize your site for mobile devices. While it may look great on a computer, it needs special formatting and coding to enhance the effectiveness on other gadgets. Studies show that more searches are conducted on smart phones than any other method. This means that the search results are also viewed on these devices. If your website does not translate well to current technology, then it will reflect poorly on you.

Don't let your website get out of date. Take the time every few months to update the data within, and, about once a year, consider redesigning the look. You might want to remove some older show references as well as adding the most recent. When you have been a part of a podcast, referenced in a review, mentioned in a magazine or other promotional items, be sure to add links to those regularly. Furthermore, you should confirm that all links on your site are working and check that your SEO strategy is still relevant.

Social Media

A great way to promote yourself and your work is through social media. Many different and ever-changing platforms allow you to engage with others around the globe. A good strategy for promotion through social media can certainly boost awareness of you and your work with others in our field. Moreover, it can be a resource for you to share original content and interact with fellow lighting enthusiasts.

However, simply posting updates about yourself and your work is not enough in today's busy world. You need to spend some time forming a network of followers, and then provide them with content that entices them to continue to refer back to you and your social activity. This content might include interesting thoughts, articles, or links to videos and websites. If you provide a source of valuable information, then others will want to follow you and your postings.

Self-promotion via social media should be planned and performed carefully to achieve the best results. Learn the various platforms and use them as intended. For instance, don't post sales specific or personal content on LinkedIn, but do share recent project photos or case studies. On every site, take the time to curate your content and make certain that your photos are interesting, clear, and nicely composed. Photos of your lighting should be a representation of your work as much as the overall show itself. Remember that everything you post is a reflection on you and your style, creativity, and originality.

Production and lighting designer Parker Genoway actively uses social media, and explains his approach:

> Social media is a great way to stay fresh on everyone's mind, even when you are not physically in front of them. Every once in a while, I throw in a picture of a show I worked on. I will get responses such as, "wow, I did not know you worked on that!" Plus it becomes a sort of resume where people can take a few seconds to scroll through my feed and see my design tastes; then make a decision off that.

When specifically thinking about social media self-promotion and the entertainment lighting field, there are certain procedures that should be followed. Most productions do not want postings of pre-production, backstage, or other candid photos without their permission. Always find out the specific rules for a show regarding content sharing. A good rule of thumb is that you can always re-share something the artist or management has shared. If they have already made it public, then you can freely re-broadcast the same content. Remember to always ensure that your photos represent not only your work well, but also the show, artist, and other designers. Above all, do not leak any secrets or special moments within the show.

Many will wait until a production has ended, or a tour has been out for several weeks before posting images. Sometimes it might just be a shot of someone at a console with the show in the background, while others might post videos of the production in progress. Remember to never post before or during a show, and rarely immediately after the show ends. Instead, wait a few weeks to a month to share your involvement. Use official event and artist tags when possible to widen your audience.

You can also make use of platforms and features to post obscure "hints" about what you are working on. Perhaps display a close-up of a console or some lights in a rig and tag the city where you are working. Be certain to not provide any identifying information about the show. This will build interest about your current project and invite people to follow you so they can learn more as it becomes available.

Some productions will require Non-Disclosure Agreements (NDA) that legally prohibit you from talking about or sharing information regarding a production. Always adhere to these rules to stay in compliance with the production's wishes. You don't want to get sued due to a violation of an NDA.

Always use honesty and clarity regarding your involvement with the lighting of a production. Do not just post a photo of a band on stage and state, "I had a great time lighting Band X." Instead, state your involvement, and give proper credit to the lighting designer and team. For instance, a better posting would state, "It was a pleasure to be on the lighting crew of Band X with lighting designer Mr Z." Honesty and transparency are principles that should be adhered to regarding all your content.

Since your individual social media represents both you and your work, keep your personal posts about yourself professional, and throw in work related images/updates from time to time. Allow people to see who you are, your interests, as well as your work abilities.

When properly achieved, social media can help you engage with others, increases your network, and might even help you land a new job or gig. Genoway was able to gain a new client via his postings and audience on social media.

Genoway explains:

> I posted a photo where I tagged both the location (the Coachella Valley Music and Arts Festival) and the artist with which I was working. The manager for another artist was scrolling through the official Coachella festival feed and found my post (due to the tag). He then

Table 11.1 Social Media Guidelines

DO	*DON'T*
Be friendly Only leave kind feedback and helpful remarks.	**Be a know-it-all** Share information, but keep it within reason. Don't respond if you don't have a helpful answer.
Post regularly At least every two weeks for work related items.	**Post everything** No one wants or needs to see what you have for dinner every night.
Be honest and transparent Give credit where credit is due and be clear about your role with every show.	**Reveal secrets** Surprise appearances, product reveals, guest artists, and favorite gags should be saved for the show's audience.
Participate Share content from peers, respond via comments, link to others.	**Set it and forget it** Platforms adapt with the times and so should you.
Connect multiple sites Link together platforms so you can post from one location. Many tools exist to facilitate this process.	**Promote only yourself** Share the work of others that you admire and give them proper credit.
Build your audience Gather many tailored followers that can contribute to your future employment and wellbeing.	**Blatantly ask for work** Posting that you need a job is like begging on the street corner.
Follow production guidelines NDAs, backstage photos, production requests, and secrecy should always be respected.	**Rely solely on social media** Many people avoid social media and if it is your only promotion outlet, you will miss them.
Understand the algorithms Know what gets featured and is likely to trend. Optimize content for the best distribution.	**Post bad imagery** Photos and videos should have good composition, lighting, focus, and interest.
Update your settings As platforms grow, adjust your settings regarding features, usage, privacy, and outreach.	**Ignore social media** With a user base in the billions, you would be ridiculous not to participate at some level.

> reached out to me because he liked the aesthetic of that particular performance, and he was curious who had been the LD. After several subsequent conversations, he asked me to submit a design for his artist.

Social media platforms are ever evolving, so be sure to keep up with the latest features, trends, and systems. Stay active and engage with others to promote yourself and your accomplishments. Remember that while a valuable self-promotion tool, many people still avoid using social media. You will run across many that never see your social media activity. It should only be one tool in your promotion toolbox.

Industry Media

Within the entertainment lighting business there are several media companies that produce magazines, blogs, websites, podcasts, and other sources of industry news. They are each hungry for content and are a great source for your self-promotion. You can reach out to any of them to share the details of a recent production or venue with which you have been involved. You can ask them to create a story around you and your work. Furthermore, you can often write a press release or case study of your accomplishments and then submit this to them for quick publication.

Early in my programming career, I would contact magazines such as *Lighting and Sound America* and *Live Design* to tell them of a show that I had recently programmed. Prior to reaching out, I would check with the lighting designer to ensure this was appropriate and also verify with the production's management regarding photos and information. Once all the approvals were in place, I would help the magazine write an article about the show and all persons and companies involved. This would put my name in the trade magazines and help promote the work I was doing. A strategy like this is still a possibility for anyone in the industry. All it takes is the time and effort to reach out to the press and let them know what is happening in your world.

Another avenue is to contact the lighting equipment manufacturers and have them write a story about your work with their products. They will then distribute this information among their customers as well as the industry press. This will often gain quicker publication and distribution as they work with the industry media on a regular basis.

Don't flood the press with everything you are doing, but be diligent about getting your name associated with productions a few times a year. This will further your reputation in the industry and might lead to new opportunities as well.

Word of Mouth

It is often said in the entertainment industry that "it is who you know," and this is certainly true for lighting. First, you must be good at what you actually do, and then whom you know will have a great factor in your career. Production managers, shop managers, lighting designers, crew chiefs, and account reps are often asked whom to hire for various jobs. When you introduce yourself to anyone, be sure to give him or her your full name and look him or her in the eye. When working on a production, take the time to meet as many people as possible. If you simply keep your head down and hang lights as instructed, no one may know who you are to ask you back for the next show.

Your reputation will precede you, so be sure to always do your best, as you never know who is watching. If you have a bad attitude and are hard to work with, then word will spread, and few jobs will come your way. However, if you are personable and hard working then you will remain busy.

Take the time to share your contact details with others, as they will likely be in situations where they need your services. All too often, I have seen technicians, programmers, assistant LDs, and others refrain from providing basic personal information with others. You never know where your next gig will come from, so make yourself known to as many people as possible. When a production or tour ends, take the time to thank each person and share contact details then too. A good follow up email after the show can go a long way and help you secure future work.

Business Cards

You should also produce business cards that state your name, phone number, email address, and website URL. This is a good thing to share when you connect with someone or when he or she asks for your contact details. The card should state something about what you do such as "lighting technician," or "lighting designer." This way when they look at the card a few days later they will remember the reference.

Never hand out your card to solicit future work while on a current production. Even if you think you can do a better job than the current programmer or LD, it is not the right place to tell the client that they should hire you next time. This mistake will always lead to the original

Figure 11.1 Some Business Card Designs are Better than Others

programmer or LD finding out and you will be seen as a job poacher. Tread lightly when handing out business cards and only give them to the client when requested.

Bring your cards with you to industry events, tradeshows, lighting shops, and any other locations where you could meet potential clients. There is nothing worse than having someone ask for your card and you don't have any with you.

Share Your Knowledge

No matter your current position or status, chances are that you have some unique experiences or knowledge that could be valuable (or at least interesting) to others. There are many resources where you can express your data, and by doing so you will also be promoting yourself. Table 11.2 details some key opportunities.

Table 11.2 Sharing Your Knowledge

Instructional Videos	YouTube is filled with videos demonstrating everything including programming, understanding DMX, mixing colors, drawing plots, and other lighting principles. Find a skill that you are comfortable teaching and make a video explaining your process.
Blog (create or contribute)	A blog is a website where people post and/or contribute writings about a particular subject. Create your own blog to share your lighting knowledge and experience, or participate on an existing blog as a contributor.
Podcast (create or interview)	A podcast is an audio (or sometimes video) recording of discussions and information. The number of entertainment lighting related podcasts is continually growing, and you could take part in any of the existing podcasts or create your own.
Learning Seminars (create or take part)	Many lighting companies host events where they invite local industry persons to attend and learn about the company and their offerings. Some of these events also look for individuals to teach lighting information. In addition, you could create your own training event, or market yourself as an expert presenter to teach your skills to theaters, production shops, unions, and schools.
Tradeshow Sessions	There are several key tradeshows within our industry such as LDI and PL&S. They are filled with learning sessions, case studies, and other teaching opportunities. Each tradeshow starts looking for people to moderate sessions many months before the show, and will entertain any idea you wish to present. This is a great opportunity to share your knowledge (and your name) with the industry.
Manufacturer Case Study	The various lighting equipment manufacturers each have their own marketing department, social media, and website. Similar to the press, they are always looking for content specific to their products. If you present them with a case study as to how you uniquely used their products, they will share the information on their website and media while also promoting you at the same time.

When employing any of the above tactics, be sure to include references to your website, email address, or other methods to contact you directly. Furthermore, use your website and social media to promote these experiences in which you are sharing your knowledge.

Socialize

Computerized communication comes in many forms and can be extremely useful for promoting yourself and your work. However, actually meeting with other people in the same business usually forms better connections. Many in our industry are willing to share their knowledge, tell stories, let you watch them work, provide advice, and more. The real key is to ask! Reach out to others that are doing the type of work you do or wish to do in the future. Meet up at industry events or create your own opportunities.

If you have goals of working on Broadway, then go to New York and ask designers, associates, and technicians to join you for a cup of coffee. Then discuss their career and explain your goals and desires. Hand them a business card and keep in touch after the meeting. Chances are that they might have an opportunity for you at some point. Even if it is not paid work, it could be the chance to sit in on a focus or programming session for a Broadway show.

The more time you spend talking with actual people in person (or even on the phone), the further your name and talents may lead to work. If others know who you are, they are more likely to hire you or suggest you for an opportunity.

Paying for Promotion

Designers and some programmers have found it beneficial to hire an agent to help them secure work. Just as a theatrical agent will find auditions for an actor, a designer's agent will work with producers, production managers, artists, and others to promote you and your work. Of course, this is not a free service, and thus they will take a percentage of the fee they negotiate for your work. Usually it works out well for both parties as you gain additional opportunities at a slight reduction in pay. In some cases, the agent will ensure you get the rate you want by just asking the client for a higher rate to cover the agent's fee. On the other hand, an agent can also negotiate for an even greater rate for you than you were expecting.

A designer agent is an excellent tool to help keep a designer working and expanding his or her career. It also lifts much of the burden from you to promote yourself, as you will have a cheerleader in your corner booking gigs for you. Some people find that agents can be limiting, however, as they may only work with a specific circle of clients or genre of production. Look for an agent who has a broad range of experience, clients, and contacts beneficial to you.

Remember that self-promotion is a key part of your job no matter your position in the entertainment lighting field. It is a full-time job that assures increased work and future career growth. As the popular saying goes, "Any press is good press," so take the time to spread the word about YOU.

12 Networking with Others

The entertainment lighting industry is largely based on relationships and thus being skilled at networking with others is very important for your career growth. If you simply do your job and no one on the crew knows your name or how to reach you, then you likely will not get many other calls for work. Whether you are an extrovert or introvert, it is important that you understand the importance of networking.

Make Contact

The real key to success in any career, besides being good at what you do, is having people know who you are and what you are capable of doing. No matter what role you have in the lighting field, you must continue to meet and socialize with others to increase your potential. This can be easy for some, yet very daunting for others. Just remember that for success, you must step out of your comfort zone and interact with others.

Start locally, talk with people you already know about lighting. Find co-workers and peers with whom you can discuss recent shows or interesting articles, concepts, and ideas. From there, branch out to people you have not yet met. Send emails to designers or programmers who live in your city and ask to meet for a chat. Even better, make a phone call to a lighting professional and tell them you want to learn more about how they have grown their career.

Many in our business travel for a living and you might discover someone important may temporarily be working in your town. If you are following a particular designer, lighting director, or crew member, keep track of when their show is coming to your area. Then reach out a few days before and see if they have any time to meet. Often there will be a day off, or even a few hours before doors, that they may appreciate meeting up with someone new. In all cases, never ask for free tickets to their show.

When you are networking, be sure to introduce yourself with your full name. Listen carefully to what the other person says and ask questions about them and their work. Do not directly ask for a job, but instead let them know that you are open to opportunities to grow your resume. Hand them a business card as a means of providing your contact details. Do not provide the card expecting work in response; it is merely a document with your contact details.

Once you make connections with others, maintain the contact with regular phone calls, texts, or emails. The best choice is a face-to-face meeting, but this is often difficult to arrange with busy schedules. The frequency of your contact with individuals should be tempered by how

Table 12.1 Conversational Skills

DO	*DON'T*
Consider body language Notice both yours and the other person's, to determine openness to the conversation and emotional responses.	**Multitask** Stay present in the conversation and away from your devices or other distractions.
Be confident Trust yourself and your knowledge.	**Interrupt** Let the other person finish speaking before you speak.
Make eye contact Maintain eye contact as much as possible. Scan across multiple people when talking with a group.	**Speak quickly** If your speaking pace is fast, they may not understand; especially those who are from other countries.
Nod Use your head to convey joy, agreement, disappointment, or sorrow.	**Cross your arms** Closed off body language will discourage others from talking with you.
Focus on the other person Keep the conversation flowing by allowing others to talk about themselves.	**Forget to give your full name** Ensure that others know who you are by name.
Listen Don't just wait to talk, instead listen, and take in what the other person is saying.	**Squirm** Stillness will confirm that you are listening and interested in the current moment.
Ask questions Start with a basic question, and then ask the other person to elaborate on his or her answer.	**Only talk about lighting** Learn more about the other person than just his or her lighting interest and occupation.
Remember names Use the other person's name when speaking with him or her.	**Share secrets** If something is supposed to be secret, keep it that way.
Be genuine Honesty and integrity from you will encourage others to share accordingly.	**Look at the floor** Stay connected to the conversation and the person in front of you.

involved they are in the relationship. Always use caution that you do not become an annoyance to someone who is not interested in sharing with you.

Look for Opportunities

Networking should not only happen with contacts that you may know or wish to know. Find industry occasions to meet new people. For example, you can attend lighting tradeshows as big as LDI, or as small as a local shop vendor showcase to make new connections. When you arrive at a trade event and visit a vendor's booth, introduce yourself, and talk to the staff about your career desires. Frequently you will gain great advice and connections. Too often, people simply look at products and do not realize that others standing in the booth may just be the key to that next big career step.

There are also other industry events such as training sessions, meet ups, and competitions. Try to attend as many as possible so you can meet the other attendees and hosts. Explain what you do and what you hope to do. Listen to their career goals and see where you might be able to work together. During industry cocktail receptions, make it a point to meet new people and talk to many during the event. Remember that it is up to you to put yourself in a position of opportunity as much as possible.

Give to Get

Networking is not a one-way street. When you are talking with peers, mentors, and others be sure to offer how you might be able to help them. Maybe you can volunteer to bring them coffee during their next rehearsal or be a light walker during a focus session. Perhaps you can help them organize their workbox in the shop. Find a way to offer assistance to the other person.

Remember that no matter your level in the field, you have some information to share. There are also many ways that you can offer your skills to help another who might not have the time or knowledge to achieve what you can provide. Ideally, networking should provide value for both parties. For instance, you might mention to a newly met LD that you recently took a course in drafting. This could lead to you drafting a plot for his or her next production.

Follow Through

After you meet someone, be sure to send him or her an email or text thanking him or her for the time. Be simple and direct and let the person know you enjoyed the interaction. Then, reach out to this person from time to time (without being annoying) to keep the interactions going.

As you develop your networking skills, you will find it easier to call others and reach out to people you do not know. It takes time, perseverance, and practice, but the rewards are immeasurable. Luckily, most people are happy to talk about their craft and often connections form that lead to additional work.

Networking should be a lifelong action that you perform at every possible occasion. With more connections, you will discover more prospects. Once you start your interactions with others, you will find it easier and easier. As with many things in life, the hardest part about networking is getting started.

13 Keep on Moving Forward

The entertainment lighting field has many options and paths that one can follow. As you navigate through your career, it will be up to you to follow and grow your own path. Remember that taking little steps leads to big changes. In order to move forward, you must continue to learn, promote yourself, and grow as an individual. Throughout life you will need to work hard on yourself in order to advance in your career.

Your Inner Voice

The lighting industry provides many people the opportunity to follow their passion to work on productions, events, and other entertainment genres. Even with a deep passion and drive, the human mind often intervenes to set one back. Your inner critic may tell you that your work is not good enough, or that you do not have the skills to do larger shows. Perhaps your mind steers you away from taking certain gigs, or discourages your taking creative risks.

Early in my lighting design career, I often questioned the positive remarks I received from collaborators about my design. I was not sure that I could trust their opinion or even their word. For instance, one director would tell the set designer that he liked the set and then complain to others about how awful he thought the set looked. How could I then take any positive reinforcement from him about my lighting with any degree of respect?

From there, I began to worry about any comments made about my lighting designs. I spoke with my father about these inner thoughts and he explained that if I had done my best, then I should be proud of my work. He also said I did not need to be concerned over other's opinions; that if I was continuing to be hired then I must be providing good lighting. I had to learn to silence my inner negative dialog and listen only for praise and constructive criticism from others.

Passion to Improve

Recently I was teaching a class when a student asked for some advice. He said that he was concerned that he was stuck in his current career position and did not know how to proceed. He felt intimidated by all the, "other people that know more than him." He was scared to take bigger gigs or high responsibility positions as he was comparing himself to others. When I asked him specifically what he was apprehensive about, he told me that he did not feel he was as skilled a programmer as others he had met.

I explained that there would always be people with more and less experience and knowledge than him. He needed to accept his current stage and sell himself at that skill level. I suggested being open with clients about his knowledge and not taking on opportunities above his existing skill set. Next, he needed to work on improving his own programming skills over most everything else in his life. If he truly wanted to improve as a programmer, then *he* had to make the decision and make it happen.

I suggested that he focus all his spare time on improving his programming knowledge and skills. This is as easy as turning off the television, ignoring social media, or not going to the bar. Instead, he should spend his time reading the console user manual, practicing on an offline editor, and going to shops or venues to borrow time on a desk. In order to fight the intimidation he was feeling, he needed to turn that energy into excitement to improve.

People with a passion do not need to be told to work harder or learn more. If you have a deep passion to work in the entertainment lighting industry, then you should apply copious amounts of your time, thoughts, and effort into doing your best and growing on multiple levels. This should also include working on yourself and your own mindset.

It seems that the human mind is often our worst critic, and thus we need to learn to direct our influences away from ourselves. This can be achieved by coming to terms with one's own thoughts. Meditation, personal growth, and therapy are efficient tools to controlling our own thoughts. Many books, videos, courses, blogs, spiritual advisors, and podcasts are available to help you dig deeper into your mind to focus on yourself and your passion.

Be Open to Change

Maybe you desire to become the next great concert lighting designer. However, along the way you find that you just don't have the interest that you once had, or perhaps you lack some of the creative or communication skills required of that position. From time to time during your career, you might need to reevaluate your goals and decide to change your path. Or perhaps opportunities will arise that take you in another direction all together. For instance, you may discover that you are better at being a gaffer for television productions than as a concert lighting designer. By following this alternate path, you could achieve a wonderful career working on amazing television productions.

Sometimes change is not easy and may not be of our own choice. I know of one lighting designer that had a rough couple of years with very little work. She had to work at various "normal" desk jobs to pay her bills. Then one day she answered the phone and was offered a job as a production manager at a large house of worship. Although it did not fit within her ideal career plan, it was a paying position within the production community. She soon found herself managing a staff of people including lighting designers and technical crew. She was able to apply her lighting designer organizational skills to her new position and carved out a great new career path.

Whenever you are accepting a new job, position, or gig, remember that you should strive to be an active architect of your own work. While many positions are clearly defined, the actual duties involved may or may not be exactly what you desire. Take the time to discuss with your clients, manager, or supervisor the expectations of your position and tell them what *you* want from the job. If you are hired as a lighting technician on a tour, you may not be able to ask for many things, but you can be clear about your skills and abilities. If you desire to become a programmer or designer, then ask for the front of house tech position (assuming you have the skills to do so). Otherwise you might get relegated to working backstage and miss opportunities to learn about programming and design.

Sometimes, a more senior person will see a potential within you and decide to provide you with a new opportunity. I know of one touring lighting technician who was offered training and a position to operate the truss automation for a tour. Although it was not a lighting position, the crew chief knew that this technician was interested in many different portions of production, and that he had the discipline it takes to operate motors over a stage. The lighting technician learned the automation system and went on to program and operate the system for a tour. Although he enjoyed the automation gig, he confirmed that he preferred lighting, and reverted back to lighting for subsequent tours. While this deviation from lighting did not result in a career path change, it did help him acquire more skills and contacts that he has since been able to apply to his lighting career.

Change is often resisted, but you should look at change as a chance for growth. Even negative change such as being fired or laid off usually leads to better things in the future. If a job is not a perfect fit, change it to better match your desires within the scope provided by the employer. Remember that only you are in charge of your own career, so make it the way you desire.

Defining Success

The definition of a successful career is a personal one. You may think success means you must be a designer for top touring acts or win certain awards. On the contrary, there are no set requirements for being successful in the entertainment lighting field. Success should be defined as, "whatever makes you happy." For some this is a perfect work/life balance, others a certain amount of money, while still others assign goals pertaining to types of shows or even winning awards.

I have seen very successful people who built a great life for themselves and their families while enjoying their work. To me this is the best marker of success. Personally, I have achieved some amazing things and still plan to accomplish more. There are others who have worked more high-profile events or earned more money, but that is not where I define my success. Decide early on your markers for success, but allow them to change throughout your life as events unfold. For instance, you may wish to design the Oscars, before you realize that the same LD has been doing it for over 30 years and is not looking to retire.

When deciding on your own personal success, remember to only compare yourself to yourself and keep your dreams within reality. Realize that results do not happen instantly, and allow time for your success to manifest within your career. Hopefully, with hard work and perseverance, you can look back at various points along your path and see the successes and growth within. Then you will truly be able to feel triumphant regarding your career choices.

14 Giving Back

The entertainment lighting industry is comprised of a rather tight-knit group of people, with many looking for opportunities to share experience, knowledge, and skills with others. When you are first starting out, you might think that you do not have much to offer, but this could not be further from the truth. There is always someone out there who can learn from you, or benefit in some way from your assistance. As you advance in your career, the opportunities to share will expand exponentially.

Teaching

The most common method that people think of to "give back" is to teach or share their acquired knowledge. This can happen via numerous different methods. One might explain to another how to operate a lighting console, or perhaps create a "how to distribute DMX" video to share online. See Table 11.2 for suggested teaching ideas.

As you teach, make sure your audience understands that you are sharing your skills and experience. When they ask questions, listen carefully. If you do not know the answer, then let them know that you will research it and get back to them. Write down the questions and then, after the training, take time to learn the answers. This then becomes your reward, as you will learn something that you may never have even thought to ask. Remember also to report back your findings to the person who originally asked the question.

Industry Charities

Around the world, there are several different charities that have been set up to assist people working in the entertainment industry. In North America, the most common for our industry is Behind the Scenes (www.behindthescenescharity.org). They provide financial support to entertainment technology professionals (not just lighting) in times of need. Grants are customized to each individual, or even their immediate dependent family, to assist with living, medical, and/or funeral costs. They also have a special fund that allows access to mental health and addiction counseling. In the UK a similar group exists called Backup (www.backuptech.uk).

As a member of the industry, you should be aware of these charities in the event that you or anyone you know ever needs their services. Furthermore, you can help the charity of your choice by making financial contributions. Each will have methodologies to make direct donations, but there are also some other ways to contribute.

One method is to pledge for a portion of your income to go directly to your charity. This will help the charity immensely, but also may be a tax-deductible item to help you reduce your tax burden. Another common way to donate is to link a charity to your Amazon account via smile.amazon.com. Once registered, Amazon will give a portion (currently five percent) of the money you spend on purchases directly to your charity of choice. This option literally costs you nothing as you were already going to give the same dollar amount to Amazon. Keep in mind that, in this case, Amazon gets the tax deduction instead of you.

Local Charities

Beyond the large industry charities, many local options are available. For instance, in Nashville lighting designers Chris Lisle and Erik Parker saw the need for a local resource for touring staff to learn life skills. In 2011, they created a non-profit charity called Touring Career Workshop (www.touringcareerworkshop.com). They host an annual free event crafted to provide all in attendance with information regarding business, taxes, insurance, retirement, relationships, and more. Sessions are presented by professionals who also donate their time by sharing their personal experience and knowledge. The Touring Career Workshop has grown over the years and now includes several events each year as well as free counseling, tax advice, life coaching, and more through their All Access Program.

This is a great example of two industry veterans who created a way of giving back that directly helps many in the industry. From online portals such as gofundme.com to regional campaigns, anyone can promote a cause and help others in need. Many unions also have charitable causes benefiting their members.

Remember that charity does not just mean donating money; it could be sharing of your time and knowledge through various resources. For example, you might offer to help a local high school with the lighting of their musical, or arrange a training class after hours at the lighting shop where you work. Charitable opportunities are abundant if you take the time to look for or create them.

More Giving

Most charities will hold various events such as raffles at tradeshows, golf tournaments, competitions, and other creative concepts to help raise additional funds. Your participation in these events (either as contributor, sponsor, or just donating your services) will be greatly appreciated. Probably the very best option with any charity (other than donating money) is to help those in need become aware of the services available to them. Share with your colleagues, talk about the charity on social media, and post signs in your shop or venue; basically, do what you can to increase the awareness to help all in our small industry.

Many designers and technicians will also take part in charity events not at all related to our industry. Perhaps you decide to donate your time for three nights programming the lighting for a cancer fundraising event. Instead of charging the producer of the event for your time, you simply complete the work at no charge. In this manner, you are using your skills to assist the event and reducing their operating costs. In turn, this allows them to raise more funds. Remember too, that you can then apply the amount you would have normally charged for those three nights as a charitable donation and possibly reduce your tax burden at the end of the year.

Mentoring

A favorite choice for many to give back to the industry is to act as a mentor for someone starting out. Many Broadway LDs will allow students and up-and-coming designers to come and sit in the theater during the pre-production of a show. Sitting quietly a few feet behind lighting designer Ken Billington, listening to how he works with his programmer, is an amazing and powerfully educational experience. No matter where you are in your career, you can always find someone with less knowledge and experience than you. Invite them to watch you work and take some time to share when possible by explaining your actions.

For instance, I was working a corporate show where my regular lighting tech had to leave after load-in for another gig. Steve, the new tech, was a young guy who had the desire to learn more about programming. During my nightly programming sessions, he would sit at FOH and watch what I was doing. He was very careful to stay out of my way, but I could tell that he had a desire to learn as I programmed. This spurred me to begin explaining my console actions to him as I continued programming. I would explain why I was selecting certain options or using particular features.

He listened intently, and asked a few questions. He was also very grateful for the opportunity, and thanked me often for sharing my knowledge with him. At the load-out, he told me that he was excited to get back to the shop and test the new concepts he had learned. I asked him to share the new ideas with others and pass it forward as well.

Mentoring can occur on-site at a production or remotely when not working a show. Perhaps you take the time to review a plot of colleague, or offer programming advice to a friend. Always be open to sharing information, as others have probably been open to helping you.

Always Ready to Share

Even if you are early in your career, you will always have some skills you can share with others. You might explain to another technician your layout for busking a show, or you might decide to host a mini-seminar in the lighting shop. Just teaching others will help you to solidify your personal skill set. Furthermore, you could post tips on Internet forums, social media, or even tweet out a new lighting trick you recently learned.

I recall a group of New York City programmers who would get together regularly at a mid-town bar. They would discuss their recent gigs as well as describe any new console functions or actions they had discovered. By sharing this information with each other, they each continued to develop their skills at a rapid pace. Furthermore, they solidified their working relationships and made new contacts for the future.

Beyond sharing knowledge, you can also pass employment opportunities on to others. When you are not able to say yes to a production, think of someone you can suggest for the job. In doing so, you will pass the gig on to another and everybody wins. This is especially important as you focus your career towards specialization. For instance, when you move from technician to programmer or designer, pass technician jobs to others who are at that point in their career.

Lighting designer Luther Frank made it a point to do exactly this when he decided to specialize his career only on programming and design. He explains:

> In addition to saying no, I found it to be a good practice to offer the work I was no longer seeking to younger professionals who were still interested and were eager for that type of work. My clients appreciated that I offered a replacement, and it helped solidify me as a designer with a contact network as well.

Share the Passion

Our industry thrives on individuals that are filled with passion for lighting. There are always new people coming into our field with a strong desire to learn as much as possible. You should always be open and ready to share your skills and knowledge with everyone. At a corporate event I designed, a motivational speaker explained that the word "passion" could be broken down to "Pass I On." She explained that if we are passionate about something, then we should share that part of ourselves with others.

The best way to do this is to share your lighting passion through your knowledge and experience. So, I challenge you to go out today and share a bit of your lighting knowledge with another, and spread your passion. Also look for ways to support industry charities and other methods to give back all that has been given to you.

15 Moving On

All good things must come to an end. This proverb has been said in many situations for thousands of years. Alas, it is true too of your career in the lighting industry. Not all people will end their career; some will just live a life of lighting until the very end. Others will choose to change careers, retire, or move into new and different roles. While planning for your long-term future is often suggested, it is not required. However, it is good to know about the choices available to lighting professionals.

A Lifetime of Light

The results of my industry survey have shown that most people work in our field for about 30 years before they begin thinking of slowing down. Generally, after 30 to 40 years of lighting, most are ready to retire or at least reduce their workload. On the other hand, a good number wish to continue working at their current pace as long as possible. It is remarkable that our industry is such a magical one that people desire to work within it for their entire lifetime.

Varying Roles

After many years working in a specific portion of the lighting industry, you may choose to change your area of specialty. Many seasoned professionals shift to consultant, teaching, venue staff, or sales roles. They will no longer directly work on productions, but rather focus their efforts on a new segment of our industry. There are many options and it can be exciting to change your daily workflow to a new, but related part of our industry.

Opportunities abound, and it is easy to transition to a new role. Often this can be accomplished while still working in your existing position. If you desire to teach or work in sales, you will likely be able to still work a few shows too. Plus, as you move from one position to another, you will bring your previous experience and contacts along with you. These invaluable resources will prove extremely beneficial both to you and your new employer.

Changing Careers

Sometimes people will find opportunities outside of lighting and change careers accordingly. They frequently find that their experience and knowledge from the lighting industry provides unique skills and insight that can be used effectively in their new field. Some opt to remain with production but focus on video, audio, special effects, or other areas. In some cases, people pivot to a totally different industry. I know of one designer who owns his own sushi restaurant and plans to focus on sushi for his future after lighting.

The small, family-like qualities of the relationships gained within the lighting industry usually help those who choose to move on to other fields. They find lighting associates are always willing to help out, and will be there should they return to lighting in the future.

Happy Retirement

Many lighting professionals choose to save money or make investments throughout their life to allow them the freedom to not work later in life. This is a great strategy because we all age and probably won't find working as enjoyable or easy when we are in our 80s or 90s.

The exact definition of retirement is deeply personal and varies widely with individuals. Some just want to live comfortably and spend time with family. Others choose to sail a boat, travel the world, or relax in the countryside. Still others will build or invest in a business that they can oversee without being involved every day.

If you desire to have a point in your life where you do not need to be concerned with an income, then it is important to start saving for retirement as early as possible. Because many in our industry work as self-employed, you may not have government savings such as Social Security available to you in the amount you require at retirement. Unfortunately, I know of several older persons who are forced to continue to work in order to maintain their desired lifestyle. Had they saved more money earlier in life, they would now be in a better place to work less (or not at all).

Retirement from lighting can be achieved and commonly designers, salespeople, technicians, and crew chiefs have a great time later in life without financial concerns. Participate in union and company retirement plans, save money, make smart investments, and support your government retirement options as well. With good planning, you can comfortably retire and remain happy for the rest of your life when *you* are ready.

Part III

The Business

No matter your position or stature in the entertainment lighting field, you will undoubtedly need to understand basic and perhaps more advanced business processes. While some are general employees, many lighting professionals find that they become their own small business and must undertake every aspect of running a company. It is important to understand the options and requirements as you move through different aspects of your career.

16 Types of Workers

The Freelancer

Many in the entertainment lighting field choose to work as a freelancer instead of as an employee of a company. There are of course pros and cons to each, and you might even do a little of both during your career. In some cases, you might be an employee at a company and a freelancer on the side. You could even become a "permalancer," where you are a freelancer continually working only with one client.

The decision to work as a freelancer should not be taken lightly as there are many things to consider when working for yourself. Some of the advantages of working as a freelancer include:

Make your own schedule: You can choose which productions you wish to work on. If you don't want to work over major holidays or certain dates, then you simply do not accept any gigs during that timeframe.

Choose your shows: Much like being in control of your schedule, you can also choose to decline certain productions. Perhaps you do not agree with the political agenda of a corporate client or you just can't stand rap

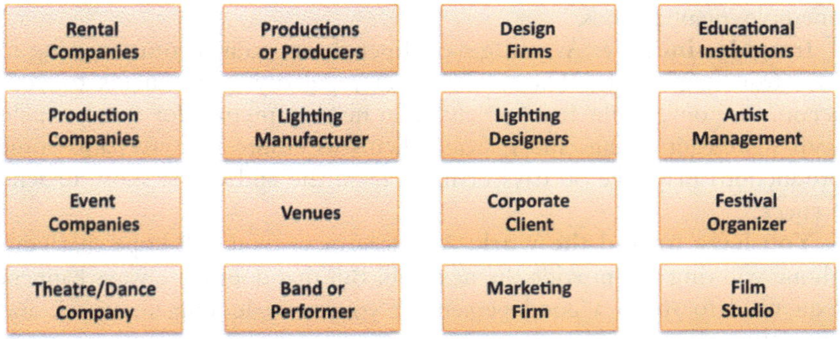

Figure 16.1 Types of Clients

music. Either way, you can turn down the gigs with which you don't want to be involved.

Easier work/life balance: When you are not working a show, you can focus on your family, friends, or pets. Not having to work in an office or meet the workload demands of an organization can be freeing.

Live where you want: Since you are not tied to a company's office, you can choose to live where you want. You may select a city based on the productions in a town, or if you are mostly touring live anyplace you wish. Many lighting designers and touring lighting staff will reside in a location with low taxes and small costs of living. Because they almost always travel to their gigs, their home location is not tied to anything work related.

You are the boss: The buck stops with you. There is no need to ask for permission, fill out annual reviews, follow corporate policy, turn in reports, etc. As a freelancer, you don't have to answer to anyone, except your clients.

Of course, there are some significant disadvantages to working as a freelancer too:

Inconsistent work: When it rains, it pours, but wow is it dry when there is no rain! Your workload may be very busy at times, and then very slow at other times. During the slow times you will wonder when you will get the next job as you watch your bank account slowly drain. Often multiple gigs will be available at the same time, yet you can only choose one. Always save extra money for dry spells.

You are the company: As a freelancer, you are responsible for all aspects of your business. You must handle bookkeeping, taxes, invoicing, marketing, training, and more. This burden can become tedious and take you away from the lighting work you love.

No benefits: When working for a company, they will often provide many benefits that a freelancer misses out on. Insurance, tax assistance, retirement, tax withholdings, unemployment, and other perks will not be available to the freelancer. Always save a portion of your income to pay your taxes. Keep an emergency fund of at least six months to cover any unemployment periods.

Irregular income: As you get paid per show, your income is going to vary from gig to gig. The amount and regularity of payments to your account is going to fluctuate wildly. You need to prepare for the dry spells with significant savings and manage clients to ensure you are paid. Some productions may take 30 days or more from the end of the show to send a check.

You have to find the work: As a freelancer, you will hope that your phone will ring and provide the next job. But that does not always happen. You need to spend a good percent of your time looking for work and promoting yourself.

Higher taxation: Depending on your location and local tax laws, you may owe more when you work as a freelancer. In the USA, an employer usually deducts payments for Medicare and Social Security. As a freelancer, you are responsible for paying into these systems yourself in addition to paying other self-employed taxes.

More complex taxation: As a freelancer, you will likely have multiple tax forms sent to you by your clients. In the USA these are commonly called 1099s and they show the government how much the client paid you. You must keep track of this paperwork as well as determine how to properly file your taxes each year.

No coworkers: When you are not working on a show, you will find that you are alone at your home office. Sure, you can call or meet up with others, but there is no one else to help you with various burdens. You can't pass tasks off to others, and you certainly will miss out on the camaraderie of company life.

Freelancing is a great way of employment and a large percentage of our industry makes a living in this manner. You need to realize that you must prepare for the freelance way of life by ensuring you always have at least six months of expenses in savings. In addition, you must actively work to promote yourself and manage your business while you are engaged in the lighting work you love. With diligence, most anyone can have a successful freelance career in the entertainment lighting industry.

The Company Employee

The alternative to working as a freelancer is to work as an employee at a company. Lighting professionals have many options available including rental/production shops, venues, cruise ships, manufacturers, television studios, corporations, houses of worship, event companies, and much more. Generally, any position in our field can find a place within a company or venue. In some cases, you might not only be an employee, but also an owner of the company.

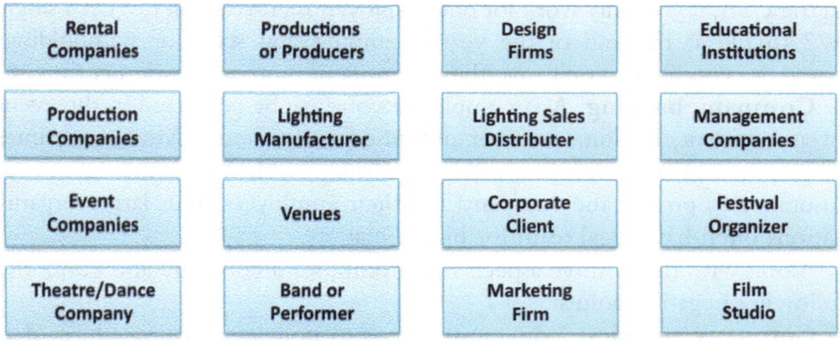

Figure 16.2 Types of Companies

As with freelancing, there are several pros and cons to working as an employee at a company. The positive aspects include the following:

Consistent income: If working as a full-time employee, you should have a reasonable idea how much your paycheck will be each pay period. Even part-time employees usually have an inclining as to how much they will be paid on a regular basis.

Benefits: Many companies will provide additional benefits to their employees beyond the employment. Health and life insurance, retirement savings programs, corporate discounts, tax assistance, bonus programs, and employee wellness are common benefits. Some employers will also provide additional dollar matching to increase your retirement savings. Should you lose your job, you may be eligible for unemployment funds.

Resources: There will be many other employees and departments that can help you with your tasks and needs. For instance, the accounting department can chase down delinquent clients, while the human resources team may assist you in personal matters.

Ladder to climb: Working for a company might provide you the opportunity to move up through various positions. Many people wish to increase their income and responsibility by changing their duties within an organization. Ayrton Lighting CEO, Chris Ferrante, started as a support technician at High End Systems and over several years climbed the ladder as a salesperson, product manager, sales manager, and eventually the general manager of High End Systems. He was then able to leverage this higher position to gain his current CEO position at a different company.

Job security: Generally speaking, you don't have to worry about whether the next show or production will occur. When working for an organization, they will have a drive to increase their business and hopefully provide you with a strong workload. If you are employed at a lighting shop or venue and there are no current shows, then perhaps you can help fix the gear or organize the shop until the next show.

Payroll: A good company will help you save by taking taxes out of your check. This helps to reduce the responsibility come tax season. For example, in the USA, if you only work for one employer, then you will just get a single W2 to file at the end of the year. Assuming you set your withholdings correctly, you likely won't owe the government more money at tax time.

Company backing: As an employee you may be protected in the event of an accident, lawsuit, or other unforeseen occurrence. Most companies possess workman's compensation, liability, travel, and other types of insurance to protect the company and their employees. Plus larger organizations often have legal team for bigger matters.

Along with the positive aspects of working for a company also come the following negative points:

Company policies: You must adhere to policies and procedures that are in place to protect and aid the company. They may or may not conform to your opinions and desires. As an employee you must comply with all guidelines. From travel policies to corporate regulations, they can assign any directive they desire.

Less freedom: When you commit to work for a company, you agree to provide a certain amount of work in return for payment. They will expect you to work required hours and days (and possibly even more hours). When you want some time off, you must ask for permission and then only take the time allotted.

You can be terminated: At any time, the company can decide they do not need your services anymore, or perhaps the company might just go out of business. Nothing is guaranteed, and no one is safe from termination.

Lower pay: Some companies will overwork employees without compensating for long hours or hard work. They may charge a client a much higher rate for your work to cover the overhead of the company. A freelancer on the same show might actually make a greater income performing the same tasks. You are often trading the company benefits and security for a lower wage.

Employees at companies can receive great benefits and job security, but company life is not for everyone. Deciding to work for another should not be taken lightly.

The Union Worker

In the United States and other parts of the world some craftspeople are organized in unions. A labor union is a group of workers who unite to make agreements about decisions affecting their work. In some cases, the members work as employees under a union contract, while others are freelancers following union agreements and guidelines.

In our industry, typically design positions are in different unions than stagehands. In either case, you could be employed by a venue or shop that requires union membership. Alternately, you might be a freelancer who the union calls to provide work opportunities, or just be a freelance union member that follows union standards and policies regarding your employment.

Some tend to look down upon unions, specifically because of their apparent aggressiveness towards work duty separations. However, joining a union can have many benefits:

Wages are regulated: The union will negotiate the rate for your work. They also set a minimum amount that you should receive for your services. This ensures that you are always making a pre-defined minimum amount per show.

Working hours are defined: The union will also negotiate for work periods, breaks, turn around minimum times, and more. They can set forth a plan that compensates you for long hours or multiple days in a row.

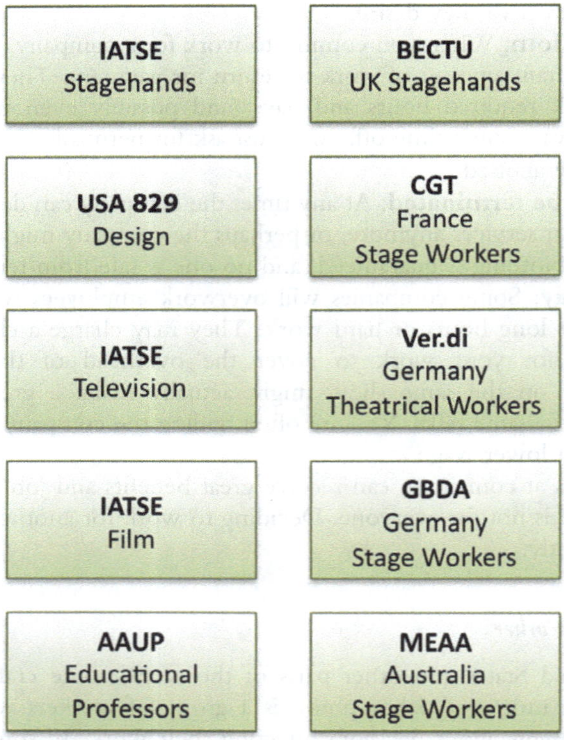

Figure 16.3 Types of Unions

Benefits are available: Most unions offer their members group insurance and retirement plans. This is a great way to help their members and a huge benefit for anyone freelancing. Additionally, unions provide compensation and liability insurance along with programs and counseling for their members. Unemployment funds may be available to union members from local governments.

Workplace safety: Poor and unsafe working conditions are never acceptable, and again the union will work with the production to ensure safety for all. If a problem arises, the union members stand together to encourage a resolution.

Payment assurance: The union will always work to guarantee their members are paid by the production. This adds a level of assurance to anyone working on shows, as payment is certain.

Per diem and overtime: Contracted rates also detail requirements for union members regarding overtime payments and per diem minimums. All clients working with union members must follow these and other guidelines.

Training opportunities: Many union offices provide training opportunities to aid their members in career advancement. In-house and external training is often provided at no additional cost to members.

Union employment does not come without its cons:

Regular dues: You will be required to pay into the union to maintain membership. Depending upon the union and location, this could be a significant amount. Failure to pay dues could lead to fines and eventual termination of membership.

Union exams and fees: In order to "get into" the union, you will likely have to pass an entrance exam and possibly pay an initial fee. The requirements vary from city to city and also depending upon the type of union and skill.

Training requirements: Many locals will not allow their members to work on productions or take specific roles until they complete union required training. In some cases, you might have to wait long periods until a class is offered or space is available for the training. This could delay your plans to start working or increasing your pay potential.

Limited opportunities: Unions tend to book employees based upon a seniority basis. This means the longer you are in the union (or have a family legacy), the more work you could be offered. Those with less seniority may find getting work through the union slow for quite a while. Furthermore, union workers may be limited to only certain venues or productions and not have access to all productions happening in a city, venue, or location.

Fee sharing: Designers often have an agreement that a percentage of their design fee goes to the union in addition to the annual dues. In exchange, certain protections, agreements, and benefits are put in place for the designer while working on a production.

Membership may be required: Depending upon your job and location, union membership may not be a choice. For instance, if you plan to work on Broadway productions, then you will need to join a union. Many television productions also require membership for designers and technicians alike.

Union workers are just as valuable as anyone else on a production. Unions provide security and support to their members and can be an asset to anyone's career. Union membership is not for everyone, and is certainly not a requirement to be successful in the entertainment lighting industry.

17 Business Structure

If you are not working as an employee of a company, then you will likely want to setup your own type of business. All freelancers are essentially already running their own business, but most are not legally structured as such. While you could continue to work as an independent contractor, you would miss out on business tax reductions, liability protection, benefits, and more. This is why many people in the industry choose to create a registered business from which to operate under.

Form an Entity

In the United States there are several different legal entities that can be created, each with its own benefits and downfalls. Some will form a Sole Proprietorship (SP), while others create a Limited Liability Company (LLC) or Corporation (inc.). In order to set up any of these entities, you will need to follow federal, state, and local laws, which may change over time. In other countries there are similar types of business structures and specific laws regarding business ownership and requirements. It is always best to get legal advice from a lawyer, accountant, and/or business manager within your country and state or providence of residence.

In addition to varying tax benefits for companies, one of the key reasons to establish a business is to gain protection for the business owner's personal assets from debt or legal liabilities. If you are working as self-employed, sole proprietor, or in a partnership, and subsequently sued for any business reason, a claimant could take your personal assets to recover a court's judgment. With an LLC or corporation only the assets of the business become available to a judgment. You certainly do not want to lose your home or personal savings due to a legal business matter.

Other benefits of setting up a company include availability of business loans, ease of hiring employees, tax deductions, and tax benefits. Furthermore, it might benefit your personal life as well because you can demonstrate that a company regularly employs you. This is usually more favorable than being a freelancer with varying income sources and amounts. For instance, continuous employment is often preferred when applying for a home mortgage, housing rental, car insurance, or personal loans. If you find that you are suddenly without income, it can often be easier to apply for unemployment from your own business than to apply as a freelancer or self-employed. Additionally, your business might have access to small business loans, government programs, and lines of credit.

There are many online businesses to assist you with setting up your own company such as LegalZoom, QuickBooks, and IncFile. Their online forms make registering a business with your state quick and easy, and the service generally costs only a few hundred dollars.

Once your business is legally registered, you can apply for an Employer ID Number (EIN) with the IRS. This free service can be accomplished at www.irs.org. You then provide your EIN to clients in place of your personal social security number. This will demonstrate to the IRS that fees were paid to your company as opposed to you as an individual. Countries other than the USA have similar systems for categorizing business tax information vs. personal tax information.

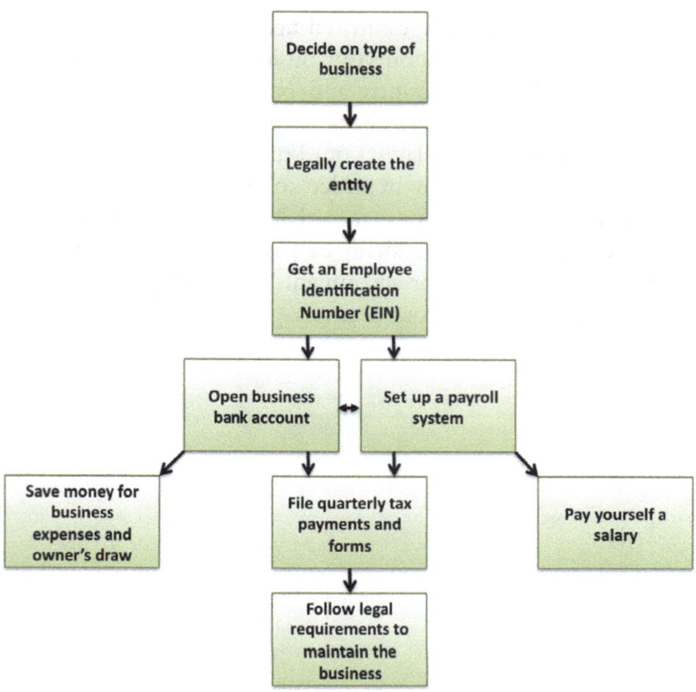

Figure 17.1 Business Processes

Managing Money

After creating a legal company, you should open a business bank account (you should do this no matter your entity type). Keeping your business income and spending separate from personal finances is a good idea for anyone, and essential with a legal company. This ensures the business dollars remain in one place, and it is easy for you (or any lawsuit judgments) to track the assets of the company. Remember that the money earned by your business is the property of the company and not your personal income. The company will need to pay you for your services, while hopefully still maintaining a profit for the owner(s) (you of course are the owner).

You will then pay yourself a salary or take owners draws (profit) from the company account as needed. Just remember that you may need to pay additional income tax on these withdrawals, so be sure to keep track of everything. Online services such as Gusto, QuickBooks, and ADP provide simple to use automated payroll resources.

Taxes and Laws

Now that you are the owner of a small business in the USA, the IRS will require you to submit quarterly estimated tax payments. Form 1040-ES and other documents (depending upon your business type) may also be required quarterly. Additionally, your country and/or state may require various payments or documents, so be sure to check your local and federal laws.

Speaking of state and federal laws, your business entity may need to adhere to certain laws and procedures in order to maintain your business status. Required bylaws, board meetings, record keeping, unemployment fillings, and other documentation will suddenly be part of your responsibilities. Again consult a business lawyer and/or accountant to confirm you are meeting the various legal requirements for your business.

18 Personal Business Paperwork

Whether you are working as a freelancer or directly interacting with clients on behalf of a company, it is important that you understand common documents and essential daily business routines. These include contracts, time sheets, invoices, and other deliverables that your client will expect from you.

Contracts

The term contract refers to a written or spoken agreement that is intended to be enforceable by law. When explaining employment, the contract should describe the scope of the work and the compensation for said work. In some cases, a company hiring you will send you their standard contract to sign. Other times you will need to send them your own contract or letter of agreement for their representative to sign. In additional situations a simple email, text, or phone call may substitute for a formal contractual document. In all cases, it is best to get agreements in writing. When negotiating, try to always be reasonable and don't make outlandish demands within your contract.

No matter the format, the contract or agreement should contain the following key elements.

- **Introductory Statement** – Your name, the client name, the project name, and dates of enforcement.
- **Terms and Conditions** – Describes the work you will provide in general terms and the amount the client will pay for that work. Also, the conditions for payment (and any deposit) should be listed.
    - You should also include details regarding travel expenses, per diem, overtime, and other clauses concerning your scope of work and the client's obligations to you.
- **Legal Disclaimer** – You may want to lay out specific legal limitations regarding your work. For instance, you might have a disclaimer on the topic of the actual rigging of a lighting plot based on your design. The goal is to provide information to the client that indemnifies you from

a broad range of consequences that a client may believe they suffered as a result of your working for them.
- **Copyright** – Depending upon your position and your deliverables, you may need to claim or release ownership of intellectual property.
- **Payment Terms** – Define when the client will pay you and at what regularity. For example, if you are working on a tour, this section should define how much you are paid per week and list the pay schedule and methodology. For a one-off event, you should include the number of days after an invoice is received that payment is expected (usually 30, 60, or 90 days).
- **Signatures** – The document should be signed and dated both by you and a representative of the client. Both parties should have a copy of the contract or agreement that includes all required signatures.

It is a good idea for you to have your own contract or letter of agreement ready to send to clients when you are accepting to work for them. Many production companies will simply sign your agreement and move forward with the booking. Others may send you their own contract and ask that you instead agree to their terms. Some productions may also use standard union contracts and ask you to comply accordingly. Remember that a contract can always be negotiated, and that you should always ask for what you think is fair.

If you are a union member, be sure that you are following union standards within all agreements, and always bring negotiation concerns to the proper union representatives. Designer specific contracts and agreements from USA829 (United Scenic Artists) are available at www.usa829.org/Contracts.

Figure 18.1 is an example of the letter of agreement I have used when programming lighting for various productions. Furthermore, a quick online search for "freelance contract template" will provide a host of samples and ideas to help you formulate your own contract or letter of agreement. Remember that a contract is used to not only to agree on payment, but also to set the scope of work and the associated boundaries.

Depending on your role, you may need to define further matters in your contract. Designers may specify first right of refusal for future productions/tours, access to artists, or copyrights regarding their design and cuing. Programmers often include a clause indemnifying them from responsibility for the rights and ownership of any video content that the production provides.

As mentioned previously, sometimes it is acceptable to work without a specific contract. If you are working as a freelancer for a production company, they may just call or email you an offer for a show or tour. If you simply reply "yes," then this can constitute a legal agreement. However, it is always a good idea to clearly state in a reply the job you will do and the amount of money you expect in return. This way if there are differences of opinion later on, you have email or text communication to

Letter of Agreement

Homer Lighting
742 Evergreen Terrace
Springfield
614-555-1212

Brad Schiller, Inc.
123 Fake Street
Austin, TX 78749
512-555-1212

Dear Homer,

Our signatures on the bottom of this letter of agreement will signify our acceptance of the terms described herein regarding the automated lighting programming of a Simpsons Sales Meeting on Jan 14-21, 2011.

Brad Schiller, Inc. will provide professional lighting programming services for Homer Lighting during a period of 8 days. In exchange for services provided, the following items for payment and reimbursement are agreed to:

- The day rate for services is $xxx for a 10 hour day. Any hours beyond the initial 10 will be charged at a rate of $xxx per hour.
- Travel days and on-site days off are $xxx each.
- Each on-site day (work, travel, or day off) also requires a per diem of $50.00 to cover meals and related expenses.
- All travel expenses (airfare, hotel, taxis, etc) are to be paid by Homer Lighting in advance or reimbursed upon receipt of the invoice.
- Travel arrangements can be made by Brad Schiller, Inc. or Homer Lighting. American Airlines is the preferred airline (business class required for international travel). Brad Schiller's Advantage number is: XXXXXX.
- Cancellation of a booking within 30 days of the event will result in a charge of 50% the day rate for the booked period (signing this agreement constitutes a booking).
- Payment is due upon completion of the event or within 30 business days of receipt of the invoice.

Brad Schiller
12/2/2010

Homer

Figure 18.1 A Sample Letter of Agreement

refer back to. When an agreement is made verbally, always send a follow up email restating the terms agreed to verbally.

Time Sheets

Many lighting professionals work on jobs paid at a specified day rate with provisions for overtime. For instance, you may agree to a daily rate for a ten-hour day (plus meal breaks). Any hours above this will be paid at a higher hourly rate. This is very common on corporate events, television productions, and some special events. It usually becomes your word against the client as to how many hours of overtime you worked. If you do not track overtime as it occurs, it will be difficult to remember after the show the exact number of overage hours you worked. In order to simplify this, keep a timesheet of your work hours to enable you to easily refer back at the time of billing. Many apps such as Clockify, Toggl, and Harvest further simplify the process of keeping track of working hours.

I prefer to use a simple Excel spreadsheet and manually enter in my hours per production. When I am provided with the initial production schedule, I will start a spreadsheet that lists the planned hours for each day. Then at the end of each working day, I will place the actual hours worked next to the scheduled time. Then in a third column I will list the number of overtime hours. Per my contract, my working time starts from the call time and ends at the time I leave the venue. In many cases, I will amend this to the time we leave for the venue and return to the hotel. Travel time does factor into your working time if the client chooses to put you at a hotel far from the venue.

You must be disciplined at maintaining your time sheet daily and ensuring it is accurate and not padded. You can also track time you spend working on the production prior to arriving on site (such as designing or programming). If payment for this time is a provision in your contract, you surely want to charge the client accordingly. While I utilize a spreadsheet, there are specialized apps and online services that can also be utilized for tracking your working hours.

Simpson's Sales Meeting				
	Schedule	**Actual**	**OT**	**Notes**
Monday June 20	Travel Day			
Tuesday June 21	8am–7pm	7:30am–6:30pm	0	meals on own
Wednesday June 22	8am–7pm	7:40am–7:20pm	1	meals on own
Thursday June 23	8am–11pm	7:40am–11:30pm	4	Dinner provided
Friday June 24	7am–11:30pm	6:40am–11:40pm	5	ALL meals provided
Saturday June 25	8am–9pm	7:40am–10:20pm	4	ALL meals provided
Sunday June 26	7:30am–10pm	7:10am–9:20pm	3	ALL meals provided
Monday June 27	7:30am–11pm	7:10am–12:20am	6	ALL meals provided
Tuesday June 28	7:30am–10pm	7:10am–8:40pm	4	ALL meals provided
Wednesday June 29	Travel Day			
		Total OT:	**27**	

Figure 18.2 Example of a Timesheet

Invoices

A very important document that you must become familiar with is the invoice. Most companies will require you to submit an invoice (especially if you are working as an independent contractor). The invoice is simply a document describing the work completed, the amount due, and the terms of payment. In short, it is a bill that you provide to the client. Your invoice does not need to be a fancy document with creative graphics; instead it should be clear and to the point as to what amount is owed and why.

Much like a contract, there are many formats of invoices to choose from, but they all share the following key points:

- **Header** – Your company name or your full name and perhaps a logo. Remember this is just a document for payment and will end up in the hands of accountants and not clients looking to hire you for your creativity.
- **Your Contact Details** – Include your name, company name, mailing address, phone number, and email address. If you have an EIN, it should also be included. Place this information on the top of your invoice, and ensure it is easy to read.
- **Client's Contact Details** – Opposite your contact details, list the client's company name, address, phone, and email. Think of this as the "to" field since it defines who is expected to pay the invoice.
- **Job Title** – Include a short title that helps the reader know what job/production this invoice is related to. Remember the accountant paying your invoice probably sees many invoices daily. A simple description such as, "Apple Launch Event September 2019" should suffice.
- **Invoice Number** – This will help you and your client identify the invoice to organize and communicate accordingly as needed.
- **Date Prepared** – This date is important as it states when you created the invoice and can be used to assist when a client takes too long to pay.
- **Payment Terms** – A due date should be specified on the invoice. It may be listed in terms (30 days) or have a specific date (October 18, 2020). The exact terms should have been agreed upon previously in your contract.
    - You may also detail any additional fees you charge if the payment is late. For instance, you could specify a late fee for non-payment by the due date. Some freelancers apply a twenty percent fee when the payment is beyond fifteen days late.
- **Breakdown of Services** – This area should be very detailed and describe the work and pricing charged to the client. You should list each day or week along with any specific overtime charges. If the

client agreed to reimburse you for additional expenses, be sure to include those here too.

  - When detailing overtime charges, always indicate the date and the exact number of hours.
  - Provide receipts for reimbursement items along with your invoice.

- **Total Amount Due** – Do not forget to have a sum total amount at the bottom of your invoice and clearly indicate this is the amount due. It should be easy for your client and their accountant to look at your invoice and see exactly how much they owe you.
- **Thanks** – Some form of gratitude or appreciation for the work is always nice to place on your invoice. Just a simple line stating, "Thank you for your business" can go a long way. You may also want to finalize the invoice with a statement such as, "Thank you, all work is complete." This makes it clear that you are no longer working on this specific project, and that no other charges will be due to your client.

Invoices can be created via a number of different methods such as on a spreadsheet, in an app, online, or in a business software platform such as QuickBooks. However you create the invoice, be sure to send it to your client as a PDF file. Alternately, you can print out the document and hand it directly to the client or even put it in the old-fashioned mailbox. Usually, however, a PDF as an email attachment is the best and most preferred method. Alternately, some apps and programs will allow you to send a link to an online invoice viewing and payment system. Always ask the client to confirm they received your invoice as even emails can be lost or skipped over.

Be sure to store copies of your invoices on your computer in an organized manner. You may need to refer back to them to confirm amounts charged to the client previously or to send the invoice again when the payment does not arrive on time.

Other Business Documentation

While contracts, timesheets, and invoices are rather common, there are other documents that may be required by your clients. They may ask to see professional proof of insurance (for equipment or workman's comp). You might also be asked to sign (or you may ask a client to sign) a non-disclosure agreement. If you are working in a design capacity, you might have certain deliverables such as conceptual descriptions, artwork, plots, cuesheets, video content, and further show specific items.

Make sure you understand the documentation that will be required of your position and organize your documents in a concise and consistent manner. Always take the time to carefully read the full text of anything you plan on signing. Ask questions about clauses you do not understand, and

Invoice

Brad Schiller, Inc.
123 Fake Street
Austin, TX 78749
EIN: 42-11111111
512-555-1212
brad@bradschiller.com

Bill To:
Homer Lighting 742 Evergreen Terrace Springfield

Date	Invoice No.
08/06/11	172

Terms	Project
Net 15	Simpsons Sales Meeting

Description	Quantity	Rate	Amount
Travel day July 28, 2011	1	$XXX.xx	$XXX.xx
Lighting Programming per day July 29-Aug 2, 2011	5	$XXX.xx	$XXX.xx
Lighting Programming Overtime July 29, 2011	2	$XXX.xx	$XXX.xx
Lighting Programming Overtime July 30, 2011	2	$XXX.xx	$XXX.xx
Lighting Programming Overtime July 31, 2011	2	$XXX.xx	$XXX.xx
Lighting Programming Overtime Auigust 1, 2011	3	$XXX.xx	$XXX.xx
Lighting Programming Overtime August 2, 2011	4	$XXX.xx	$XXX.xx
Travel day August 3, 2011	1	$XXX.xx	$XXX.xx
Per diem for daily meals and related incidental expenses	7	50.00	350.00
Off-site airport parking in Austin, Texas	7	8.00	56.00
Taxi to Hotel		62.59	62.59
Taxi to Airport		61.94	61.94
Thank you for your business.		Total	$1,000,000.00

Figure 18.3 Sample Invoice

never sign something with which you do not agree. If you have concerns about a contract or agreement, refer to a lawyer or union representative for assistance. No job is worth signing something you do not agree with or completely understand.

19 Business Operations

Every position within our industry must adhere to some basic business practices. Most people work in some sort of capacity that could result in salary discrepancies, contract disagreements, cancelations, and other business operations. It is important to understand the basic principles and procedures to ensure smooth sailing through your career.

Getting Paid

At some point during your career, you will have a client or company that delays, forgets, or refuses to pay you for completed work. You have several options when it comes to collecting your money. These range from simply following up to taking them to court. It is always a good idea to start with the kindest method first.

First confirm that you sent your invoice, timesheet, and other documentation to the correct person and that their payment is late per the agreed payment terms. Looking back at your contract, invoice, and emails should easily provide you these details. If everything looks good on your end, send your client a gentle email reminding them of the payment due and ask for a date they intend to pay.

If a customer continues to evade payment, reach out to the principal owner or business manager. Moving up the chain of command can often increase awareness and create action. Always be polite and use direct phone communication when possible. Clearly state the facts and ask for a quick resolution. After the conversation, follow up with an email to ensure you have a paper trail.

If non-payment continues, you will need to prepare to take the client to court. Before hiring a lawyer and entering costly court fees, send a certified letter to the company letting them know your intentions to file a lawsuit based on the non-payment. Often you will find this is enough for them to finally send your money. Otherwise you will need to decide if the court costs are worth the hassle of suing versus writing the situation off as a loss.

Although the goal is to have the client pay you the agreed amount, in some cases the best resolution may be to offer a settlement. Suggest a payment plan or a reduced total payment. Some payment is always better than non-payment.

Pre-Payment Options

In certain situations, you may wish to negotiate pre-payment to ensure you receive the money you are promised. This can be very important with a new client or with a production that you think might be less secure than other employers.

For instance, I was hired as a programmer for an electronic music festival back before major corporations ran them. The producer was just some dude that used to produce raves. I did not have lots of faith in his ability to pay me, or even be located, once the show was over. So I added further clauses to my contact requiring him to pay 50 percent of my fee 30 days before the event and the other 50 percent before the first act hit the stage. He promptly paid 30 days in advance, but his tune began to change on show day regarding the rest of my money. He kept promising to send a check the following week. I actually ended up in the production trailer with LDs from other venues at the festival as bands were entering the stages. We all insisted on cash payment before we would go back and turn on the lights. Finally, he gave us each the cash he owed us, and we went back to our consoles and provided a great show.

Pre-payment plans are certainly reasonable forms of business. In some cases, you may ask for early payment due to client concerns, or perhaps to simply help you financially before the event. Many designers ask for a portion of their fees to enable them to earn an income while designing concepts and drawing plots. If you do plan to request early payments, always be sure to include this information in your initial contract or agreement.

Settling Contract Disputes

In some cases a client may not follow the guidelines outlined in your contract. Your process of resolving these disputes is very similar to that of non-payment. First, point out their error, and ask for a correction to the situation. If they continue to not comply, then move up the chain of command. Always remain firm, but kind. Do not resort to yelling or threatening statements.

Hopefully with reasonable discussions you can work out your dispute. In other cases, more drastic actions may be required. You may even need to resort to a lawsuit, but be sure to take all other reasonable actions first. With certain situations you could take your grievance to social media and speak out against the contract offender. Again, be sure to remain honest and kind while clearly presenting your grievance. Reserve this step as a last resort as once your information is online it cannot be taken back.

In some cases, you can leverage your yet-to-be provided work against their non-agreement to your terms. For example, a lighting director friend of mine worked for a major band for nearly twenty years. For all of those years his contract has stated that international travel shall be booked in business class. The band's management had always complied. On a recent tour the band decided they no longer wanted to pay any crew's business class ticket. It did not matter that the band was still top of the charts and flying in private jets. When my friend found out that they would book him coach for a twelve-hour flight he refused to travel. The tour manager asked him to "take one for the team" and sit in coach. Instead, he held his ground, and continued to state it was unfair and against his signed contract. As the date of travel loomed closer, the tour manager was finally able to gain clearance for him to fly business. The band eventually agreed that it was more important to have their usual lighting director than to go against this particular contract point.

Cancelations

Events, shows, tours, or any gig can cancel suddenly for multiple reasons. It is always essential to have a clause in your contract regarding these situations. Once you book a production you will not be able to take any other work on those dates. If the production suddenly cancels, then you will have missed out on other opportunities. For this reason, most agreements in our field have a statement similar to, "Cancelation of this booking within thirty days of the event will result in a charge of fifty percent of the rate for the booked period."

When a show cancels within your defined window, you should promptly send an invoice for the agreed amount and clearly state it as a cancelation fee. Most customers will honor your contract and follow through with payment. If you are working on a tour or recurring production then you may need to alter the language of your contract accordingly to protect yourself regarding the period of time for lost work. For instance, if the lead singer in a band loses his voice and the tour ends two weeks early, you want to have a clause that ensures you receive proper payment during that downtime.

When working for a union or as an employee for a company, be sure to follow the guidelines of your employer. They may require specific actions when a production cancels and you need to adhere to their standards. Always check the union or corporate policies regarding legalities and contracts.

20 Tax Form Basics

First, please remember that I am not a tax accountant or taxation expert. Keep in mind that that your country or state may have different policies and laws than mine. However, it is important to understand taxation basics, as they exist in the United States. You can then apply this knowledge to your local laws and procedures.

Paying taxes is a required part of life in most countries and should never be taken lightly. Be sure to consult an expert and always pay your taxes. Learn your local laws and procedures and follow them explicitly to avoid fines, fees, and possible jail time.

Self-Employed Taxation

If you are self-employed, then the companies that contract you will have you fill out a form W-9 for US federal taxes. This is a basic form that simply provides your client with details such as your social security number or taxpayer ID and other personal information. They will use this data to report to the government how the amount of money they paid you.

Then, at the end of the year, the company will send you and the IRS a form 1099 that shows the IRS how much they paid you within that specific year. It is your responsibility to pay your taxes due on this amount. Remember, too, that just because your employer does not send you a 1099, this does not mean that you can skip paying taxes on income you received. Always pay tax on all received income, even if you do not receive a 1099! In the past a form 1099-MISC has been used for all non-employees. Starting in 2020, form 1099-NEC (Non Employee Compensation) has been reinstated.

In most cases, it is advised (and maybe required) to pay quarterly tax estimates so that you do not have to pay the full tax burden at the end of the year. If you are working through 1099s without any tax withholding, be sure that you are saving or prepaying an estimate as to what you will owe. You do not want to spend all your earned money and then have a large tax bill with no cash to pay for it!

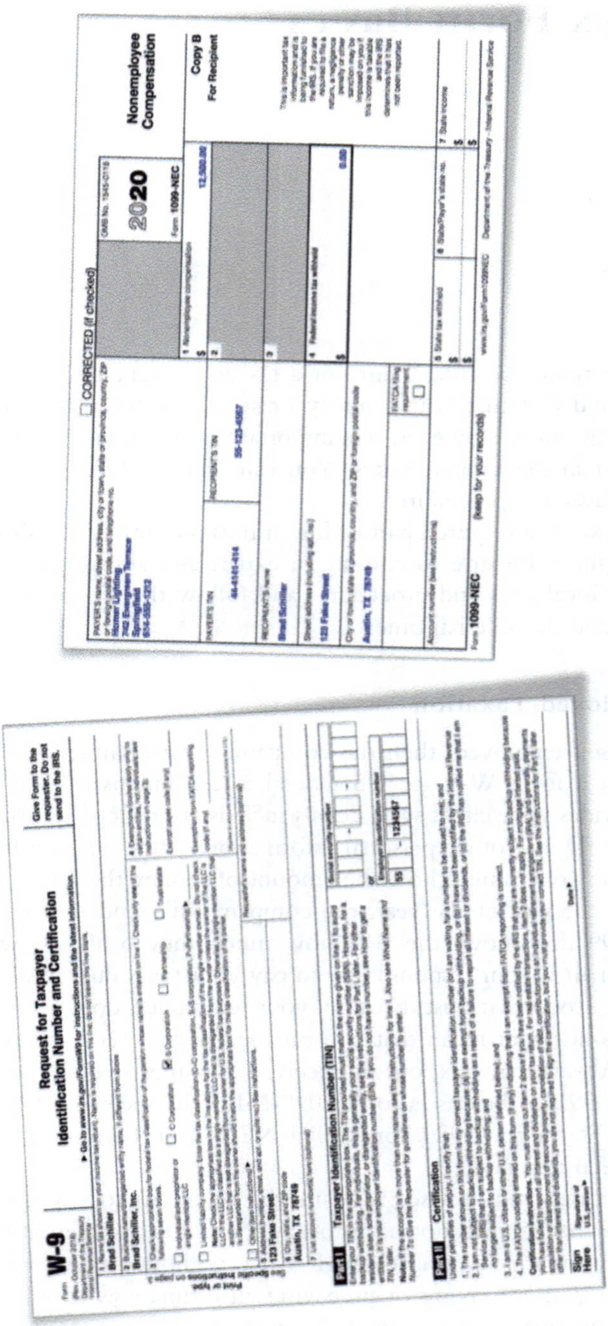

Figure 20.1 IRS Forms W-9 and 1099-NEC

Company Employee Taxation

If you are working as an employee of a company, then you will fill out a form W-4 when you are hired. This tells the company how much money you wish to have withheld from your paycheck for tax purposes. You will answer a series of questions regarding marriage, dependents, and additional deductions on this form.

Then, at the end of the year, your employer will send you a form W-2 that indicates how much money was contributed to the federal, state, and Social Security/Medicare taxes. Depending on the amount withheld, you may owe additional tax funds, or be entitled to a refund. Regularly review your W-4 selections and make changes as needed for optimum tax withholdings.

Pay Your Taxes

No matter your methods of withholdings, you will need to ensure that you properly pay your tax burden(s) by the required dates. This can often be difficult when you are touring the world, although e-filing has certainly made it easier. Generally, you will need to fill out forms by entering data contained within all your W-2 and 1099 forms. Then follow the procedures to explain the amount of income you received and how much money you withheld (prepaid). You can also subtract certain deduction amounts as determined by current tax laws. Any differences in the final taxation numbers will result in either you owing the government, or them owing you.

This is a very simplified version as to how to begin to think about your taxation. The laws are ever changing, and there are many things that you might be able to deduct to reduce your tax burden. For instance, if you are self-employed or own your own business and purchase specialized tools for your work, some or the entire amount paid may be eligible to reduce the total tax owed. Services such as websites, Internet, cell phone, accountant fees, and equipment are usually legitimate business deductions. The acceptable line starts to blur around home offices, mileage, meals, and other life vs. business expenses. Always follow the law and consult tax professionals for assistance to ensure you are adhering to allowed deduction procedures.

Help Is Available

The majority of people working in the lighting industry hire the services of a Certified Public Accountant (CPA). This professional will fill out your tax documents for you and possibly act as a business consultant regarding taxation and financial decision-making. A CPA should be up to date with all the federal, state, and local tax laws in order to ensure you meet the legal requirements while also looking after your best financial interest. Most in

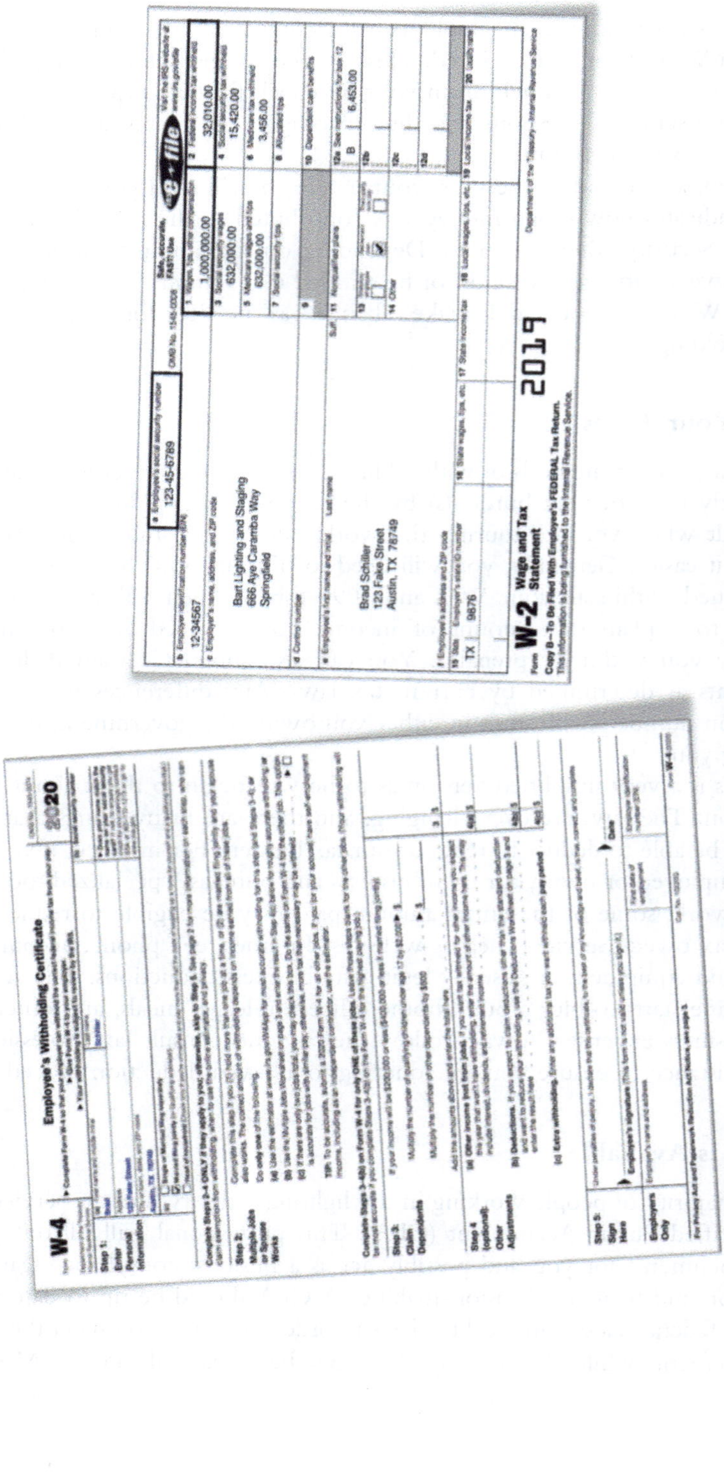

Figure 20.2 IRS Forms W-4 and W-2

our field find it best to hire a CPA familiar with the entertainment industry. While you likely will not find a "lighting CPA," you can enquire with your local peers as to who they hire. Furthermore, some accountants specialize in working within the entertainment industry, helping musicians, actors, photographers, and other similar professions. They are always a good choice as they better understand the cycles and procedures of working in show business.

The IRS provides a wealth of information and forms for freelancers, employees, and business owners on their website at www.irs.gov. In fact, they have a dedicated section titled Small Business and Self Employed Tax Center, which can be found at www.irs.gov/businesses/small-businesses-self-employed. This site is a one-stop shop for all related documents, procedures, and information regarding the IRS and your business.

IRS Resources

The IRS has numerous documents available on various topics. The following are key IRS documents with which you should become familiar:

IRS Publication 334 – Tax Guide for Small Business
IRS Publication 463 – Travel, Entertainment, Gift, and Car Expenses
IRS Publication 505 – Tax Withholding and Estimated Taxes
IRS Publication 535 – Business Expenses
IRS Publication 583 – Starting a Business and Keeping Records
IRS Publication 587 – Business Use of Your Home
IRS Publication 936 – Home Mortgage Interest Deduction
IRS Publication 946 – How to Depreciate Property

Basic Documentation

It is essential that you track your income as well as outgoing expenses. No matter your business structure, this will be beneficial when taxes are due. In addition, it can help you realize if you are financially successful with your current employment choice.

When working as a freelancer, company owner, or union member you will want to keep as much of your business finances as possible separate from your personal funds. The best way is to have a business banking account and a business credit card. With these tools you can keep money separate from your personal needs and expenses.

Next, you should track every dollar going in and out of these business accounts through a software program or app such as QuickBooks, Xero, FreshBooks, or FreeAgent. These tools allow you to document all your finances in one easy location. They even generate reports that can be easily sent to your CPA. Other tracking tools such as mileage logs, receipt scanners, and project tracking will come in handy for every freelancer.

With all this documented information, it will be much easier to claim tax dedications and reimbursements where allowed and to prove the validity of your data should you get audited.

When a company employs you (as opposed to you being a freelancer), then you will receive your paycheck and not need anything further from a business financial reporting point of view. However, your company may ask for expense tracking and receipt scanning. If you are allowed to make purchases on behalf of a company, then you must understand the Travel & Entertainment (T&E) policies of the company, and learn their expense reporting tools. These systems are usually in place so that the company has the data required for their own taxation needs. Failure to follow their rules may result in you paying out of your own pocket for business expenses.

It is best to have a separate credit card specific to these business expenses; thus many companies will issue you a card when you are hired. Be sure to get reimbursed for all allowed business expenses, and never pay out of your pocket for a valid charge that the company should pay. Honestly report your expenses, and always spend the company's money as you would if it were your own. Keep in mind that the expense reporting process is primarily to aid the company with deductions at tax time.

Taxation is a required result of earning money for everyone. Throughout your career, always pay your taxes, and ensure you understand the basics of taxation in your country. When needed, hire professionals to help with forms, requirements, and documentation. Remember, too, that as your income grows, so too will your tax responsibilities.

21 Getting Gigs

One of the most common questions people ask when starting out in the entertainment lighting industry is, "How do I get more work?" Some of the best advice is to work hard, be nice, and talk with others. If people know who you are and what you are capable of, then there is a chance that they might hire you. Of course, you need to have some skills and knowledge about the field as well.

Get Personal

Email, social media, texting, and submitting resumes are great, but this is a business of relationships and you must actually meet people face to face. Take the time to make phone calls, arrange meetings, and truly meet with professionals in our industry. If they have looked you in the eyes, there is a better chance that they will think of you for future work.

Networking is a key skill that will aid your career as much as lighting skills and knowledge. You could be an extrovert and find networking simple, or an introvert struggling to connect with people. Many introverted personalities find it difficult to start conversations, but it is something you must do to gain career opportunities.

When you find yourself talking with other lighting professionals, management, producers, or artists, be curious and listen. Ask questions and do not focus on selling yourself. Most importantly, follow up via email or a phone call. As you openly discuss lighting and life, you will find that opportunities may present themselves naturally.

For example, I was recently speaking at an event when a young person came up to me after my presentation. He asked some general questions about lighting as well as specific questions related to my session. I then asked him what he was interested in and where he was currently working. He shared that he liked lighting, but video content was his area of specialization. He explained and showed me examples on his computer of his work. I was very impressed with his conversation and examples, so mentioned I had a need for some custom content on an upcoming show. After a few more interactions, I hired him to create and provide some content for my show. Had he never started a conversation with me, he would have missed out on this chance to expand and sell his work.

Not all conversations will lead to immediate employment, but the lighting industry is rather small, and you never know from where your next job will come. Furthermore, you can gain important knowledge from your connections with others that prove valuable for your future.

Get Out There

Just because you create a portfolio and/or website, this does not guarantee future work. You must get your name and skills in front of the people making hiring decisions. This will differ depending upon the exact position you seek, but the basics of marketing yourself remain the same. First, realize that if you do not have much experience, or are relatively unknown, most people will not take a chance with you in a high position. You must earn their trust and respect regarding your knowledge and abilities, and this only comes through proven hard work.

If you want to work more in concert touring, then seek out the lighting companies that are providing crews for tours. Get to know their staff, hiring managers, and crew chiefs. You may have to work in the shop for a bit to prove yourself, but it will be time well spent. Let them know your personal goals (designer, director, programmer, crew chief) and take gigs that support your future. You might have to meet with them on several occasions, take them for coffee, or call repeatedly to remind them of your desire to work with them. Do not be too hasty though, as that could come across in a negative manner.

Perhaps you want to be a lighting designer for television or theater. If so, then going to lighting shops is probably not the best place to start (unless you have zero experience as a designer). Instead, you need to associate yourself with producers, theater companies, television studios, and other designers. Again, you must determine who the people making the hiring decisions are and start networking with them. If these people know who you are and understand your abilities and experience, then they are more likely to think of you when a show needs a designer.

Whatever you desire for your future, there are jobs out there for you. You will need to research all the performing arts centers, houses of worship, production companies, lighting shops, television studios, and other consumers of lighting and then make contact with them. Opportunities abound in every city in the world, but the first step is simply getting potential employers to know who you are and what you can offer them.

Rejection Happens

Sometimes it can be frustrating to talk with people over and over when looking for work. Similar to dating relationships, there is often more rejection than acceptance. However, you will eventually find companies and events that wish to hire you. Until then, you must accept the rejections and continue pressing forward.

I once mailed out (before email was common) a hundred copies of my resume looking for design work in the Dallas, Texas area. I blanketed the area's theaters, comedy clubs, production companies, theater groups, and venues. I thought that I would receive many offers for work. Instead as the weeks went by, I became rather disappointed that no one was reaching out to me. What I failed to realize at the time was that there were three key principles working against me:

1. Jobs were not available right then, just because *I* was looking.
2. My work experience was very limited.
3. I did not follow up.

I just sat by the phone expecting work to come to me. Instead, I should have called each of the locations and asked them if they received my resume. I could have spoken with someone and further explained who I was and what I could do for their organization. Even with my limited design experience, there were likely some companies that could make use of my skills. I never spoke with anyone directly to explain this, and instead let only my resume speak for me.

Months later, the phone did ring, and I received one design job from that exercise. Had I followed up, I probably could have achieved several more. Remember that shows occurring soon have likely already hired a production staff. You probably won't get put on a show in any position for some time after your initial contact with an organization. The timing of your employment query might dictate that there are no productions needing your talent at that particular moment. Jobs only materialize when organizations have a need for you and your talent.

Rejection can happen for many different reasons. Don't let it get you down; instead use rejection as a lesson to learn how to improve for next time.

Get More Work

There is no easy answer to the question, "How do I get more work?" The best choice is to be a hard worker and make sure people know who you are and your capabilities. Then network with everyone possible to become known in the circles within which you desire recognition. Don't be in a rush, and be sure to always follow up with others. With these basic skills, the jobs will come your way.

22 Planning for Your Future

At the beginning of their careers, most people are not ready to think about retirement. It is usually 30 to 40 years away, and the income of a new job is exciting to put towards current needs. However, it is essential that you plan ahead in some fashion for your future. Inevitably, there will be a time in your life that you wish to work less, or not at all. Proper financial planning now helps you live comfortably later without income concerns when you slow down or retire completely from working.

Generally speaking, most experts and financial companies agree that you should save approximately ten to fifteen percent of your annual income for retirement. In theory, it is fairly easy to stockpile money for later on. For instance, if you simply invest $200 a month for 48 years, you will amass over a million dollars. This calculation assumes a seven percent average annualized growth in a tax-deferred account. Forty-eight years is a long time, of course, but if you are currently in your 20s you will likely be employable for the next 40 plus years. Additionally, if you increase the amount saved per month, then the time duration to save a million dollars will decrease dramatically.

While saving $200 a month sounds like a great plan, it usually is not realistic for people starting out. If you have significant debt such as student loans, rent or mortgage, car payments, credit card bills, and insurance then you might not have an extra $200 a month to save. Furthermore, if the savings interest is less than your debt interest, then you will not be helping your financial situation by stashing extra cash away. In many instances, it is best to pay off major debt first, before investing in your retirement. Remember, though, that it never hurts to start saving as early as possible.

The Best Path to Wealth

The very best way to ensure financial security, both in the present and the future, is to earn more and spend less. As you earn money throughout your life, you may desire bright, shiny, new items. In America, much of society is focused on material objects and ownership. Do your best to avoid reckless spending, and avoid debt as much as possible. Never purchase things you cannot afford, and live within your means as much as possible.

As your career grows, you should strive to increase your earning potential and actual income. If you have been a lighting technician for five years, making the same amount of money per week, then it is time to increase your rate. Look for opportunities throughout your career to boost your income. Usually you will have to ask for an increase, as employers are always looking for the best deal they can get when paying employees. Take responsibility for your own income and ask for a raise or promotion every few years.

I once worked for an employer who did not offer raises and preferred to keep employees at their same salary year after year. I worked for this employer for quite some time, increasing my knowledge and value, yet they did not financially compensate me for becoming a better employee than when they hired me. Furthermore, due to economic inflation, the salary I was paid was actually worth less than when I started. A smart friend pointed out that I was actually being paid less and less each year (in terms of the dollar value). I finally went to the HR department and asked if they thought I was worth less to them now than when I started. I also asked for a raise and negotiated for "cost of living" increases moving forward.

Take the time to ensure you are being paid a fair wage for your current role, and as you grow in your career ask for increased compensation for your services. Over your lifetime, your income should increase steadily and provide you with more opportunities to save for your retirement. Try not to spend all your newfound income, and always be diligent regarding your savings plans.

Actual Retirement Savings

When it comes to saving for retirement, there are many choices available to those of us living in the United States. Many other countries have similar options available and it is important to research the possibilities within your country. Furthermore, the assistance of a financial advisor will certainly prove beneficial towards any retirement plan.

As an Employee

If you are the employee of a company or a union member, then you probably have some sort of retirement plan available to you. You can elect to have a percentage of your income deducted from your paycheck and placed in an investment of your choice. Many 401(k) plans allow you to select the type of investment for your retirement dollars. Take the time to study the options and select a strong choice. Furthermore, the money going towards retirement savings is usually tax-deferred, meaning you won't pay taxes on this income until you withdraw it at retirement age.

The amount you contribute per check is up to you, but generally it is a good option to save ten percent or more of your income.

If your employer offers a matching program, always collect as much of this as possible. It is free money that is yours by simply electing to save a specified percentage of your income for your future. For instance, assume your employer will match 100 percent of the first three percent you invest. If you are earning a salary of $60,000 this means they will add $1,800 into your retirement savings annually (assuming you also save at least three percent). Remember that there are few regulations regarding matching, and employers can create these programs as they wish. Some will match within certain constrains or caps, and most will also require vesting. Often you must be employed for a certain period of time before the matched dollars actually become yours. If you leave the company before this timeframe, then you will not keep the matched investment dollars. Always take time to fully understand the retirement program offered by your employer and ask questions of the HR team.

If you change employers, transfer your 401(k) dollars from the previous employer to your new employer's plan. There are many rules and regulations regarding transferring of retirement funds, so be sure to educate yourself and get help as needed. Improper transitioning of plans could result in fines and/or taxes due.

As an Individual

If you are working as a freelancer or own a small company, you still need to save for retirement. Obviously, an employer based 401(k) will not be available, but you do have several good options for creating your own retirement savings plan.

Individual Retirement Account (IRA)

An individual retirement account (IRA) is available to anyone who has an income, whether self-employed or working for a company. The government sets and adjusts contribution limits. Currently if you are under 50 you can contribute up to $6,000 a year to a Roth or traditional IRA. If you are over 50, then you can contribute $1,000 more annually. There are two types of IRAs, Roth and traditional. The key difference between the two is that a Roth is taxed annually, and a traditional IRA is taxed when withdrawn. Contributions to an IRA are tax-deductible and will help to reduce your tax burden at the end of the year.

When you are younger and earning less, it is a good idea to invest in a Roth IRA as you will be taxed at your current tax bracket. In the future, you will hopefully have greater income and thus be in a higher tax bracket. Then a traditional IRA benefits as the tax is deferred until retirement.

Simplified Employee Pension (SEP)

Anyone working as a freelance employee is eligible for a Simplified Employee Pension (SEP) IRA. This investment option allows you to contribute up to 25 percent of your freelance income or $56,000 a year; whichever is less. You can also fund it up until April 15 of the following tax year. This enables you to determine how much you wish to save after you have completed and calculated your income for the previous year. Also, with a SEP IRA, you are not required to make a contribution every year, which can be helpful when life happens and you wish to spend your dollars immediately instead of saving for retirement.

Solo 401(k)

As the name suggests, this is a 401(k) plan for an individual. There are limits as to how much you can contribute individually, but you can also make contributions as an employer. So you can increase the individual $18,500 annual limit to $56,000 depending upon IRS rules and your income. Solo 401(k) plans are set up as either pre- or post-tax and often include additional fees from investment companies as well as complex IRS paperwork requirements.

Even More Options

Beyond traditional funds and government programs, there are additional ways to save for your future. Real estate, businesses, stock markets, certificates of deposit, and other investments can yield great returns. I know of many in our business that own multiple rental properties. Not only are they earning monthly income from the renters, they also hope to receive a hefty return on the sale of the property in the future. Others have invested money in various business ventures including lighting companies, rehearsal studios, restaurants, car washes, and equipment.

Diversifying your savings with various types of investments is usually a good idea as you will average out better if something goes wrong with one of your choices. Don't go into debt to form an investment, rather take the time to save money first and then invest wisely.

Government Programs

The Social Security program in the United States allows workers to save money for retirement. At the age of 62 you become eligible for retirement benefits and can begin receiving a monthly stipend from the government. Other countries have similar programs where residents contribute a portion of their income to the government in exchange for benefits after a certain age. While most programs are required by law, some are also optional.

Always check your local laws and procedures to ensure you are in compliance and eligible to receive benefits when available.

A Helpful Resource

At least one company is based upon providing financial education and resources to employees in the arts and entertainment industry. The Institute for Financial Wellness for the Arts (www.theifwa.com) has training courses, calculators, information portals, and links to financial coaches. Their services are available to both individuals and companies. They can assist you in your retirement planning as well as daily financial decisions. Several major financial companies such as Fidelity Investments, Charles Schwab, and Vanguard provide retirement planning and assistance without a focus on the entertainment industry.

Have a Great Life

Throughout your career, your retirement needs, choices, and abilities are likely to change. It is vital that you continually plan for your future regarding retirement savings. Periodically, you should evaluate your plan and decide if it still suits your perceived needs. As you move between employers or to and from freelance positions, always keep saving. Talk with a financial advisor about your best investment options, and limit your spending to only within your means.

With a focus on your future, you can live comfortably now while also saving for later in life. Unfortunately, some in the entertainment lighting industry have lived a very successful life, and yet have not planned well for retirement. This has left them struggling later in life. On the other hand, many colleagues have saved, and will be very comfortable for their eternity. Choose now for a better life by saving for your later years. With a solid financial backing it will be easier to decide when to slow down or retire from working.

Part IV

The Creativity

Entertainment lighting provides for many fun and exciting careers. Whether you are involved as technician, designer, programmer, or other position related to entertainment lighting you will find your creativity beneficial to your craft.

Many people are born with a natural talent for creativity, while others have to work at it. The lighting industry is unique as it allows you to utilize both your technical and creative talents. The very core of a lighting designer's job is highly creative, but he or she must also possess technical skills to develop and communicate the plot, concepts, and looks desired for the show. A lighting programmer provides creative input as looks are built on stage by selecting colors, positions, movements, and effects as they relate to the production's needs. However, the programmer's job of entering data into a desk is also a very technical position. A lighting technician utilizes technical skills to repair and operate lighting equipment while creativity adds expertise to fixture layouts and uses. By combining your technical and creative skills you can achieve exciting and fun lighting for productions.

If you don't feed and exercise your body regularly, then you become a fat slob who tires easily. In the same manner, if you do not frequently feed and exercise the creative portions of your brain then you will become a non-creative, boring artist who fails easily. Lighting is an art and art requires innovation and expressive freedom that only comes from new thinking and ideas.

Each position in entertainment lighting provides many avenues for creativity. However, it is also easy to rest on your laurels and continually reuse the same concepts, procedures, and ideas. Instead, free your mind from the traps of stagnant thought and become more creative as you approach each production with new understanding and insight.

23 Abundant Creativity

Each role within the entertainment lighting field has creativity to contribute. From the lighting designer to the shop staff, there are opportunities to be thoughtful in new and distinctive ways that can greatly enhance a production or routine.

Artsy Lighting Design

A lighting designer is hired by the artist, director, or producer to craft the creative look of a production's lighting. Of course, the lighting also needs to be functional, allowing the audience to see the event occurring on stage. But beyond illumination, lighting can provide additional creative dialog for a show. Color choices, movement, positions, and shadows can convey emotion, attitude, and time, as well as other concepts. The lighting designer must determine how to best use entertainment lighting to enhance the production without compromising budget, time, and other valuable resources. To be successful, a lighting designer must be creative and innovative throughout his or her career. Only by approaching the same problem with new solutions can a designer truly provide a production with a creative use of entertainment lighting.

Creative Programming

Lighting programmers are frequently the driving force of entertainment lighting. They command the lights on a daily basis and must be familiar with not only the abilities of the fixtures and console, but also the subtleties of design on the stage. As a programmer it is very important to be creative while maintaining the technical accuracy required when building looks on stage. Quite often a lighting designer will ask the programmer to create cues on stage, and then the LD will suggest modifications. For these reasons, the automated lighting programmer must be a creatively minded individual. A purely technical programmer is nothing more than a data entry monkey with little to no creative input in a production.

However, there is a fine line where the programmer's creativity must not override that of the lighting designer. A programmer must always remember that the LD is hired as the responsible party regarding the overall look of the lighting on a show. Many times, a programmer may be asked to create a look or sequence that is not pleasing to him or her, but is desired by the LD. The programmer must also determine how best to maintain an open relationship with an LD, without compromising his or her own artistic abilities. This proves a difficult task for many, and thus why you will frequently see an LD working with the same programmer across multiple productions.

Motivated Technicians

The entertainment lighting technician plays a highly important role in a production by installing and maintaining fixtures. Without proper equipment maintenance, the production will suffer, as the audience will not properly see the creative efforts of the designer and programmer. The tireless work of technicians ensures that productions are repeatable for audiences, day in and day out. One might question how a technician can provide creative concepts or "think outside the box" to help a production; however, these actions are essential to the success of a production. Technicians regularly contribute artistic ideas for fixture placement, rigging, and usage. A designer may not possess awareness of certain limitations or abilities that only a technician can achieve through specialized knowledge of the fixtures and the specific lighting rig. Technicians should speak up and suggest new methods for entertainment lighting practices to further enhance the creativity on a production.

Let's All Work Together

Entertainment lighting within a production requires many skilled craftspeople to fully achieve a production's needs. Only by working together to provide the best possible show can all the expertise unite. Open sharing of ideas helps to inspire others to produce unique solutions and innovative concepts. Everyone involved with a production needs to feel free to contribute ideas and thoughts. Even if a specific concept is not utilized, it might inspire further notions that ultimately benefit the show.

Speak up and share your ideas with others and see what transpires. Creativity sparked from one person will often ignite a passion in others. Whatever your position, verbalize freely your creative ideas and see where they take you. Of course, you don't want to criticize the efforts of others as you make suggestions, so use caution with your phrasing. Once ideas are on the table, be open to modifications or suggestions and always keep the best interest of the show in mind. When we all work together, we can help create the best show ever for every production.

24 Creative Inspiration

Creative inspiration can come from various sources, not just from our minds or co-workers. The creative aspect of working with entertainment lighting involves continuous study of countless different resources and processes.

Show Biz

The most obvious place to look for the creative process of entertainment lighting is in other productions. I find it very important to watch many different types of productions and to try to remove my "lighting specialist" hat. That is, I try to watch from a *normal* audience member point of view. This has not been easy, of course, because I naturally want to dissect the technical aspects of the show. Over time, I have gotten better at simply watching the entirety of a show without continually looking at the lighting rig.

DVDs, streaming media, and televised events are good resources for viewing shows, but remember it's best to see a production in the medium in which it was originally created. The lighting in concert tour DVD or video stream always looks very different from viewing the live concert in person. I suggest watching shows of all different sizes and types. Go see the big shows in Las Vegas, on Broadway, or popular concert tours, but also check out a local high school production or community theater show. You will see that limited resources often enhance the overall creative output. Also remember that ticket fees are often tax deductible as research expenses.

One of the most stimulating productions I saw in recent years was awe-inspiring due to its sheer simplicity. While on vacation at a small beach resort in Jamaica, my family and I attended a nighttime "circus" act on the resort grounds. Walking up, I noticed they had a small four-foot high pipe-and-base setup with four PAR cans, each with a different color gel. These were connected to a small wooden plank with an on/off switch for each of the four fixtures. As the show started, I watched the lighting guy as well as the circus acts. He actually sat right at the base of the lights and changed the focus and turned on or off the different fixtures to create various cues. He was the board operator and "automation" for the lights. He repositioned the fixtures during blackouts, as well as created live "followspot" moves. These four fixtures represented the *entire* lighting rig for this show. In no time I focused myself back on the show and found it very enjoyable. Suddenly it did not matter if they had one hundred moving lights, or just the four PAR cans. The lighting cues fit the production and were well thought out in a creative and unique way.

I was very inspired by this performance, and on several productions after the event I found myself lighting more-with-less in certain cues or moments. This is not to say that I did not do this previously, but suddenly I had a new creative impulse to do so.

Natural Stimulation

One of my favorite inspirations for lighting comes from nature itself. Even though I may not try to replicate the exact look of a sunset or lightning strike, I find the randomness and beauty in nature more inspirational for my own creativity than any other source. I love to look at the colors of light in the early morning as they are projected on the side of my house or through trees. How come I can never get an automated light to make that perfect sunrise golden-amber color? Yes, they can get close, but there is something about that natural color that fails to be perfectly replicated with technology.

Figure 24.1 Nature Provides Many Sources of Creative Inspiration

Shadows, textures, reflections, clouds, patterns, colors, shapes, movements, and so forth are all readily available to be studied just outside your window or venue doors. Simply spend a little bit of time studying the environment around you and noticing the intricate details. You will be amazed at your increase in creative potential during your next lighting session.

Sometimes I watch the representation of leaves from a tree as the shadows dance around the sidewalk and think about how to replicate this on-stage. My eventual replication may or may not be a direct recreation of that moment. For example, I might use a completely different pattern at a much faster speed during a rock song, but try to emulate the random undulating feeling of that shadow play. Or I might gain insight about turbulence and randomness from watching waves break on a beach.

When flying on airplanes (which I seem to do often), I like to get a window seat and just stare out the window, day or night. I find viewing the Earth from this perspective unique, and I actually look forward to examining the atmosphere and terrain. Again, I may not directly replicate what I saw out the window, but I know thinking about how different our planet looks from 30,000 feet often inspires my lighting.

Artsy-Fartsy Assistance

Another great method to boost creative juices is to observe various types of artwork. From paintings to photographs and sculpture, the art world is based upon creativity. Again, it is best to view the artwork via the intended medium. While you can view many paintings in a book or online, nothing beats seeing the actual works of Vincent Van Gogh or Salvador Dali in person. Given the fact that the Great Masters offer us a large spectrum of work to view, there are many wonderful galleries located in every city of the world. Nowhere is there a shortage of opportunities to view artwork.

I even carry small postcards in my workbag of several works of art that have inspired me in the past. I will glance through these from time to time to remind me of how I felt upon viewing the originals.

I like to look deep into artwork to understand the colors, layout, composition, textures, and inspiration that contribute to the uniqueness of the piece. I even find the brush strokes of a painting or lens choice of a photograph often relate to selecting a movement or speed for a lighting cue.

There are also a host of ingenious artists who choose to use light as their primary medium. As they are working within the same realm, they are a great resource to gain inspiration. James Turrell, Olafur Eliasson, Anthony McCall, and Dan Flavin each have amazing bodies of work. Anyone interested in lighting should become familiar with their art and view it whenever possible. Their artwork can be found in installations around the world and make a fabulous destination on a day off or during a vacation. You can simply Google each by name to learn more about their work and installations.

Creative Writing

I recently did a search for books on amazon.com for the word "creativity," and found over 80,000 titles. While they all may not be winners, there are many wonderful books regarding triggering and exercising your creative talents. In fact, you are reading a great source of creative inspiration at this very moment!

Many years ago, lighting designer Jules Fisher told me that he reads *A Whack on the Side of the Head* by Roger von Oech once a year. I have this wonderful book, and while I do not read it that often, I do try to read it every few years. The simple exercises modestly point out that we should not get trapped into the same ideas and methods, and instead should "whack" ourselves upside our heads to help develop new ideas and concepts.

Creativity Is Just One Talent

As stated before, you can enjoy entertainment lighting due to both its technical and creative aspects. Creativity comes in many forms and can be applied no matter your current position within the lighting field. Throughout your life, you have likely heard the term "think outside the box." Only by actively improving your creativity, and applying different lessons to your work, can you truly become successful at providing innovative lighting and solutions.

Part V

The Lifestyle

Working in the entertainment lighting industry is often a career of passion more than a career of financial wealth. While not producing as many millionaires as other professions, the field does allow workers to have a wonderful career, nice salary, and great work/life balance. Although there might be difficulties related to relationships and health, working in entertainment lighting is generally a wonderful choice.

25 Why We Do It

The majority of the people working in the entertainment lighting industry are doing so to satisfy an internal passion and not to earn millions of dollars. If your desire is earning heaps of money, then this field is probably not for you. However, if you want to really enjoy your work and make a good living, then you have come to the right place.

Lighting people usually start off working as technical crew in high school or they come from being a musician in a band. The majority find a calling towards lighting, and many feel that "lighting is either in your blood or it is not." The joy of being a part of a production, helping to illuminate actors, augmenting a band's performance, and bringing emotion through imagery are only the start of what drives us. We get to work with amazing people, state of the art technology, travel the world, and help entertain others. No one lighting person has ever created an entire show alone, and being part of a larger team is very satisfying on many levels.

Some want to work as artists and find the art of lighting matches their creative desires. Others just enjoy the technical aspects, or the thrill of bringing entertainment value to audiences. A few even satisfy their dreams of being a performer, but achieve it in a technical method rather than focusing on a personal performance skill.

Backstage magic happens endlessly around the world and many find opportunities to be a part of the crew from various positions very enticing. Lighting enhances productions, allows followers to see their idols, and entertains directly with colorful beams and movement. Whether working as a manufacturer's salesperson, lighting designer, or lighting technician, each person plays a pivotal role in the success of a production.

A very good financial living is possible working within the lighting field, no matter your position. You do not need to become a top lighting designer to live a very comfortable life. Many find the lighting field a profitable environment from which to live within their desires. While only a few at the top can make millions of dollars, most within our industry are able to afford all they want as well as save for their retirement. Having a quality life while doing something you love is a true marker of success.

Work Hard, Play Hard

Putting a show together involves lots of hard work, both physical and mental. People in all positions and levels of our industry work very hard to bring lighting to productions. If you have the passion, then you will find joy in the hard work, and not find mundane tasks a burden. Many people find they are in the zone (or flow state) when doing their work with lighting. I believe this to be a key factor in a successful career.

Often, when working on a drawing, programming for a show, or simply loading in or out, I lose track of time and stay present with the duties in front of me. This mindset is commonly referred to as a flow state and is often achieved by people who are totally engrossed in the moment. Experts agree that the condition is good for our minds and wellbeing. Therefore, the more we can achieve this frame of mind, the happier we will become. Whenever you enter the flow state during your primary work, then you know that you are following an internal passion.

You must enjoy the majority of the work that you are doing; otherwise it just becomes labor. Sure, you might not want to have to roll up that FOH snake after an outdoor festival, but you do it anyways. The inner passion to be a part of the production allows you to gather the strength and desire to get your hands dirty and roll the cable. Not every single task within lighting will be fun, or even easy for you.

Besides, at the end of the load-out you know that your friends are waiting for you. By friends, I mean the rest of the lighting crew. Most lighting people stick together and actually enjoy spending time with each other. Like-minded people tend to enjoy the company of similar types of people. Those working together on the lighting of a show create many friendships and working bonds. Our industry is based upon relationships, and most of these are formed and solidified during working actual events.

Limited Stress

Generally, our industry provides mostly low-stress positions of employment. Sure, there can be major stress to get a show's lighting finished on time and on budget, but we are not trying to save human lives. We are working within the entertainment industry. It is very important to remember that most productions have a limited run and thus will be over at some point. New tours, productions, and products will be created, and old ones will be forgotten.

However, stress does occur regularly and is usually related to relationships with others, production politics, or limited time. Learn to recognize stressful situations and try to minimize your personal pressure level increasing. Take a moment to think about the situation and respond instead of reacting. When an irritant occurs, don't react; always wait a few seconds or minutes for a better response to come to mind. This will lead to smoother conversations, improved workflows, and lowered stress.

When in difficult circumstances, I will often sing to myself the first few lines of the song, "There's No Business Like Show Business." This calms me down and reminds me of the joy and purpose surrounding the work in which I am currently involved. In fact, take a moment to look up the full lyrics of the song to gain further show business inspiration.

Because the stress in our field is limited and production based, most enjoy their careers with relatively low personal work-related trauma. If you can learn to see the stressful situations as fleeting, then you can start anew with the next production. There are enough stressors in our personal lives that we do not need more from our careers. Luckily, the entertainment lighting field provides generally low-stress jobs for most. By releasing the burdens of previous shows and managing your time, you can lead a low-tension life.

Unbelievable Situations

We are very fortunate to work in an industry that is built upon helping people find joy and happiness. The entertainment field affords lighting professionals many opportunities that are not always available to those working in "normal" jobs. Often, we experience some of the high life as we work alongside celebrities. Furthermore, the very nature of the entertainment lifestyle brings us into locations and situations that may seem unbelievable to outsiders.

Tour buses, private planes, five-star hotels, exclusive venues, fine food and drink, private performances, and unique locales can be yours when working in the entertainment lighting industry. I have exited a limo and walked the red carpet at the Oscars, flown on a band member's private Lear jet, slept in presidential suites, been escorted by police motorcades, visited top-secret sites, and experienced many more incredible occasions. These exciting moments may not come early in your career, but they are certainly possible, no matter your position or role. These occurrences abound and will happen when least expected. Furthermore, lighting happens all over the world, thus paid-for travel to any place on the planet is also possible (and very common).

When traveling for work, try to take some time to actually see the location you have arrived within. Sure, there will be instances when you only see the venue, but make the most of your time and get out and see the city whenever possible. You can see quite a few sites in a town over the course of just a few hours. When you can, schedule an extra day to explore the locale to which you have been sent. If you are on a production and have a day off, get out of your hotel room and see something in the local area. Why would you ever skip going to see the Eiffel Tower when someone else has paid for your trip to Paris?

Lighting is Life

Natural light drives much of the life on our planet. The plants and trees could not live without sunlight, and lighting allows safety within our communities. Working with this amazing medium is very satisfying. A career within the entertainment lighting field will provide you with a good financial lifestyle as well as amazing opportunities that you never thought possible. The rush that you feel when working on a large production while thinking back to your high school beginnings is unimaginable. If lighting is in your blood, then you will automatically strive to do all you can to grow your career in a way that suits you best.

26 Work Life Balance

We all want to live a balanced, healthy life. However, the desire to grow and participate in our career can often take a toll on other aspects of our life. Our lives consist of more than just our work; our lives also comprise important facets such as family, community, spirituality, and social life. Learning to balance your work with all of these aspects of your livelihood and adapt to life's situations is vital.

Most workers in any industry might easily fall into situations where their work detracts from their personal life resulting in many depressing and devastating situations. Unfortunately, the lighting industry is not immune from this type of pitfall and often requires more time away from your family and personal life than common professions. It is very important to find a way to balance your work and your personal life, even if you feel your work *is* your life.

Working Hours

First, realize that the entertainment industry generally works during the hours that others are looking for enjoyment. Quite often, this means working nights, weekends, holidays, and other untraditional hours. This will have an impact on your social, spiritual, and familial lifestyles. Interacting with your friends and loved ones will become more difficult as they plan events and gatherings that conflict with your work schedule.

Lighting departments are often asked to work when others from the production are not present. This is because you are turning the lights on and off and making it difficult for carpenters, performers, band members, and other crew to accomplish what they need. Therefore, you are usually regulated to working the overnight hours after the others have completed their time on stage. Then, you must restore to the hours of the production's rehearsals and shows. Adjusting to these shifts in your working hours can take heavy tolls on your mind and body, as well as be confusing to your family and friends.

Remember too that the show is going to happen with or without you. You are just one part of the entire production. Your needs to schedule around personal events just won't matter to the overall production. Instead, you will generally need to schedule your life activities around the needs of the show. Asking for a night off to take your spouse to the movies does not happen in our field. More than likely, you will need to arrange your personal events according to the free time in the production's schedule.

Growing Your Career

When working as an independent in the lighting industry, you will want to say yes to as many opportunities as possible. Often this will mean missing life events such as birthdays, holidays, celebrations, and social activities. If you are offered the chance to design the lighting for a producer you have been courting, you probably will not turn it down just because of a single night's personal conflict. Instead, you will take the gig and miss that particular occasion in your life.

As your career grows, it should become easier to selectively say no to conflicting events, but this is not usually the case. Most build a work ethic around saying yes to gigs above all else. This routine then develops into difficulty when making decisions about life events versus work opportunities. Most people will default to work choices and let their social and family life suffer. However, there are many times when it will be more important to select your personal life over career moments. You need to weigh each situation carefully and make informed choices. There will always be more gigs, but many personal events won't ever happen again.

I Love My Work

As stated earlier in this book, most enter our field with a strong passion for working with light. This leads to a drive to work more shows because the work does not feel like labor. It just feels like part of life and generally is lots of fun. This outlook leads to a wonderful lifestyle as you are not dreading working. However, it can also lead to confusion and depression when it comes to choosing your passion over your personal life.

There are certainly times in your life when you should focus a large part of your energy on your work. However, as relationships foster and families grow, the responsibilities outside of work increase exponentially. The decision-making process will become more difficult. Leading a life consisting only of lighting work is not a complete life, even if you absolutely love what you are doing.

Finding the Balance

Having a wonderful, well-rounded life means living with a balance between your professional and personal lives. Believe me, it requires effort and discipline along with understanding the people in your life. You need openness and honesty regarding conversations with your clients, co-workers, family, and friends to maintain the balance.

Whether traveling or just working wacky hours, there will be many situations where you need to choose between employment and personal moments. Missing life's activities such as birthdays, school events, social commitments, and personal time so that you can earn money to support yourself and your family is a very commendable mission. Nevertheless, you must also weigh the emotional toll on yourself and others when choosing work over other parts of your life.

I was once programming for a lighting designer when he made a phone call while sitting with me at front of house. He called his young son and wished him a happy birthday. I heard him apologize for again not being there, and then he asked if his son had received the gift he had shipped him. It became clear that his son was more concerned with his father's absence than wanting to look at the gift. When the call was over, the designer was noticeably shaken and needed time to recover from the call.

In that moment, I thought of my own son who was only a year old at the time. I decided right then and there that I did not ever want to miss his birthday. I resolved that I would do all I could to ensure that I was never in the same situation as this designer. I made a vow to myself that I would not work on his birthday until my son was old enough to say, "Dad, I am doing my own thing for my birthday." For the next fourteen years or so, I was always present for his birthday celebration and not away working. I was lucky that this meant turning down only one gig, as the rest of the time I found ways to ensure I was home. I am forever grateful to that designer for teaching me this lesson and motivating me to stick to my convictions.

Personal discipline is key to a good work/life balance. You must check in with yourself from time to time to ensure that you are happy with the choices you are making. If you feel that you are missing too many social events, then make a change and turn down a show, or ask for some time off. Although the entertainment business works many long strange hours, it is also flexible when it comes to life. After turning down a show for personal reasons, you will find that more opportunities will come your way. Don't let the drive for more money or experience take away from life's precious moments with those you love, including yourself.

Relationships

Personal relationships, marriages, and family connections are the most common parts of our lives that must be balanced with our work. Open and frequent communication is very important, and luckily modern technology makes this a bit easier to achieve than in the past. To find a balance, communicate with your loved ones often when away working on projects. Take the time when walking to catering to call your child or spouse.

Instead of going to the bar after the show, choose to FaceTime with your parents. It is too easy to simply focus on our jobs and put our family aside until we return. But it is far more important to keep our family in the loop with our lives than to ignore them.

Maintaining a love relationship is very difficult under many circumstances, and our industry requirements certainly do not make it any easier. We are often traveling, working in environments of excitement, and meeting many people. Those who remain at home are often unclear as to what exactly goes on backstage and form their own thoughts. Some of the stereotypical temptations are real and others are imaginary, but that does not make them any less a concern. Couples must take the time to be transparent about their lives with each other and to correspond honestly. When a partner lets you know that they need you home, then make time to be home. A good companion will also understand your desire to work more and provide you the freedom to work as much as you wish.

Unfortunately, many love relationships in our industry do not survive very long due to the rigors of the road. Usually there are numerous other factors that contribute as well, but a big ripple in most comes from time apart. Do all you can to balance your absences with phone calls, texts, emails, and other methods of comfort. Never take your loved ones for granted and always let them know you are thinking of them. Small tokens of sentiment go a long way during times of absence.

Life Happens

Throughout your career, both good and bad circumstances will occur in your life. Some will affect your career directly and others will transpire with little to no impact. When life's situations do take place, be clear and honest with those you work with and ask for the help you need. Everyone experiences deaths, family events, celebrations, personal troubles, and relationship concerns. If you are working a show and get unexpected news, do not be afraid to let the production know what happened and indicate that you need to leave. Our industry is very understanding and helpful in these occurrences.

For instance, I was on a tour and my father was selected to receive a very important honor. I really wanted to be at the ceremony to share this momentous occasion with him. I spoke with the production manager of the tour and discovered that the event date conflicted with only one show during the tour. Furthermore, the tour stop was a festival day. Subsequently, I was allowed to fly to my dad's event after the previous show and return to the tour after his award ceremony. I simply had to teach someone on the crew to take my place at the console in my absence for the festival show. I already had a backup person, so this was easy to achieve. While I was exhausted from the extra travel, my family was thrilled I was able to make the event and the tour easily continued during my absence.

I have heard of similar stories for sudden unfortunate events too. Trust me that no show is worth staying at if it means missing important life moments. Death and accidents happen throughout everyone's life, and in most cases you need to choose to be with your loved ones during these times. You cannot get missed time back, and the show will always go on, with or without you. With open dialog among all concerned, you should be able to balance your career responsibilities along with life's events.

Take Care of YOU

The best way to find a healthy work/life balance is to check in with yourself as much as possible. If you feel you are missing out on personal events, self-practices, relationships, social engagement, or other parts of life, then dial down your employment responsibilities. Have open discussions with loved ones and co-workers to help you maintain a balance. Even reject work from time to time to focus on other aspects of your life. There will always be more jobs, but many of life's occurrences won't ever happen again.

If you feel that life is getting out of control and that balance cannot be achieved, then please look for some assistance. There are many ways to grow personally including books, seminars, therapists, coaches, and charities. Help is available, and you simply need to decide to start improving your life. A good career does not only involve working with light, it is also important to have a balanced life with meaningful relationships, social activities, and strong connections. Be sure to balance yourself often to ensure a well-rounded life and career.

27 Allure of Celebrity

Working in the entertainment industry allows many interactions with celebrities of all types. I have had the pleasure of meeting rock stars, actors, models, CEOs, politicians, and even an astronaut who walked on the moon. Celebrities come in all types, and working with them presents some unique situations and challenges for those of us working alongside them. But what, or who exactly is a celebrity? An online search defines the word celebrity as, "a famous or well-known person." Anyone from a musician to a business icon can be a celebrity. When working within the entertainment lighting industry, you will undoubtedly work near and alongside several celebrities. No matter your role, it is important to understand the characteristics and expectations of working with famous people.

Working with the Famous

I am often asked by non-industry friends if I get to meet famous people and what it is like to work with them. I think they expect me to give some sort of tabloid gossip to further explain the behaviors of a rock star; however, this is not what I share. I usually explain that working with a well-known person is no different than working with a not-so-well-known person. We are each an important person on this planet. Just because millions adore one person, and another is known by only a few, should not make a difference.

When a "star" comes to a production meeting or visits FOH, I treat him or her the same as any of the other professionals I am working with. I respect that since they are the "artist" they can make certain decisions, but this is no different than an office staff having a meeting with their CEO. Quite often the celebrity is a very creative individual who has some interesting ideas to share. His or her years of entertainment experience come from a unique perspective that often leads to inventive concepts for lighting looks and staging. It is important to treat these celebrities as individuals, and not show them preferential treatment or considerations. In other words, don't kiss their ass just because they are famous. They get plenty of that on a daily basis, and most just want to cooperate to get the show to a perfect state.

Respect Their Privacy

I happen to have the cell phone numbers and email addresses of some of the celebrities with whom I have worked. This personal information was provided in trust based on our working relationship. I respect their privacy and do not contact them to simply chat about the weather or impress my friends. I am merely a working professional that has been a part of the same productions as them. We are not buddies, so I don't feel I need to call any of them just because I know their number. When we are working on a project and need to speak about the show, then I will make contact; otherwise I refrain from arbitrary communications.

Furthermore, your website and social media should not contain photos of you hanging out with famous people. You are not in their company as a fan. You just happened to have worked on productions that included a famous person or two. I have seen some lighting staff who post photos of themselves socializing with celebrities, only to later get in trouble career-wise due to their flagrant lack of respect. Even photos of celebrities without you in the shot could be controversial. I was recently visiting a show and one of the crew members showed me photos of the show that he took from his unique backstage location. He said that he wanted to post some on his website, but had to get permission first from the management. That was the right attitude, as his angle and perspective were different than those of all the audience members with smart phone cameras. Perhaps the artist or production did not want to reveal imagery from this other position.

Autographs are for Fans

If you have ever been on a tour, you probably learned the unwritten rule about the toilets on the bus. Another unwritten rule in the entertainment industry is to *never* ask for autographs. You would not ask the lighting tech for his autograph, so why would you ask that of the person performing on stage?

Sometimes you might get an autograph as a gift, but you should never ask. I remember when I had a check for a show that was personally signed by Oprah Winfrey. When I went to the bank, the tellers could not believe that I would rather have my money than keep the signed check. I insisted that I wanted to be paid for my work and preferred the money over the signature. The teller reluctantly cashed my check and also provided me with a copy of the check so I could keep the autograph. However, I gave her back the copy, as she was more thrilled about the signature than me.

Some artists will be very friendly with the staff and may offer to sign items for you. Be sure to allow this to happen naturally and openly. If you manage to get signed memorabilia, don't run out and sell it. This is a personal gift to you that should be cherished.

Keep Your Cool

Remember that celebrities have many people who want to talk with them. If you are at a show and need to have a discussion with the artist, you might have to wait a bit. Sometimes artists will call you into the dressing room before or after the show to discuss the evening's performance. Remember that they have very busy schedules and thus you might need to accommodate their timing needs.

Don't think that just because you are involved with the lighting that you have an open invitation to socialize with them. They have many other obligations as well, and while your dialogue is very important, try to remember they are being pulled in many directions at once. When you do chat with celebrities, keep it professional and always bear in mind they are the "top of the food chain" regarding the performance. However, do not let their status compromise your professionalism or the safety of yourself or others. Feel free to inform them when they are wrong or an idea is dangerous or inconceivable.

Lighting and production designer Butch Allen has encountered many celebrities over his long career. His interactions have ranged from brushes with famous folks when working as a lighting technician or lighting director to detailed collaborations with stars as production designer. Always the professional, Allen shares the following about celebrities:

> I've been very fortunate both professionally and personally to meet quite a few people some would consider to be celebrities. Sometimes it's a "sweaty hands, words fail me" moment, and other times the first encounter is an "old friends getting together" very easy kind of kinship. After twenty plus years in this business I still get the chills from a true "brush with fame" moment.

We are all human and of course will find times that we are personally thrilled to meet certain famous people. I will never forget the moment that I shook the hand of Buzz Aldrin while working with him at a corporate event. Take these moments in stride and remember that your career sometimes awards you with these special moments.

We Are All Celebrities

As I stated at the beginning of this chapter, being famous just means "well-known." Consider too that there are many different types of celebrities. "Famous" is not reserved for worldwide public recognition. You might get a chance to work with a well-known lighting designer or production designer and feel more butterflies in your stomach than when you work with a movie star. Perhaps this LD is a bigger celebrity to you any movie star, and that is okay.

Whether a big-name LD or a world-renowned pop star, every person should be treated merely as another professional coworker. Sometimes they will become friends, and sometimes they will be a thorn in our side. Either way, famous people are commonly an important part of our lives when we work in the entertainment industry.

28 Healthy Living

"But what if I die?" was the question I heard the LD ask when I told him to go back to the hotel and recover. At some point in our careers we will experience working with ill people or being sick ourselves during a show situation. Unfortunately, the age-old credo "the show must go on" often takes precedence over our own wellbeing. Of course, no production is worth dying for, and in the same light we must consider what benefits we really provide to a show when we are feeling under the weather.

Health Matters

Hours of late nights, high stress, poor eating habits, and dirty conditions will lead most humans to some state of illness. Although there are many preventative measures that usually help, illness is bound to eventually happen. I was working a theatrical production where the LD was not feeling well all through load-in and pre-production. He was overworked, overstressed, hungry, and sleep deprived. During the lighting focus he nearly fainted a few times while looking up at the rig.

I decided that enough was enough and suggested that he return to his hotel room to get some rest. I suspected that if he could rest for a few days that he would be in great shape by show time. First, I rounded up the master electrician, producer, and lighting crew chief to explain the situation to all of them. Then they helped me to assure the LD that we could cover all that needed to happen in the next 48 hours without him. Only with total moral support from his team did he feel comfortable enough to get some rest.

He actually slept for over 30 hours straight! After some rest and medication he was able to return for the final rehearsal and the show opening. Of course, we had all pitched in to assist while he was out, and he was pleasantly surprised by the amount of work that had been accomplished without him.

Sometimes you need to help out a co-worker and let them know when to throw in the towel. Frequently when sick and overworked, it can be difficult for a person to choose his or her own health as a priority.

Sick as a Dog

When touring, there is usually no backup person ready to fill in running a console for a single night. Due to calling spots and operating a desk there are very few people who can instantly take over a show at the last minute. Unfortunately, this means that the touring lighting director often has to run the show, no matter what his or her condition.

I witnessed an amazing display of this one evening while touring with Metallica. On the tour, Butch Allen would run one desk and call spots, while I would run cuelists on another desk. One day he was extremely sick and stayed on the bus as long as possible (looking and feeling awful). As show time approached, he mustered up enough strength to come to FOH and fulfill his duties to the best of his ability.

Although his playback that evening was nowhere near as outstanding as normal, he did manage to call all the spot cues with near perfection. I helped by taking over as much of his console operation as possible so he could concentrate mostly on the spots. Often, during the show, he would call to the spots, lean over the FOH railing, and puke his guts out. Then he would wipe himself off and turn on his headset mic again in time for the next spot call! As soon as the show was over, he hurriedly returned to his bunk on the bus to sleep. This rest was imperative to help his body heal as quickly as possible.

I don't recommend operating shows in his condition, but usually there is little choice in the matter while on the road. Butch did the best he could by resting and medicating all day so that he could use all his strength during the two-hour show. Because he communicated about his condition, he also had me and other staff assisting him as needed.

Going to the Hospital

Of course, there are rare times when a person is so sick they cannot continue with a production. A friend of mine was working with a production as a consultant for their digital lighting. On the first day the LD/programmer said he was feeling bad and he needed to go to his hotel room to rest. The next morning, everyone discovered that the LD had been so sick that he checked into a local hospital overnight. Suddenly, my friend moved up from consultant to LD/programmer and found himself staring at a console with which he was unfamiliar. However, he quickly took on the tasks and helped the show move forward.

The LD made the right choice by deciding that his health was more important than a simple corporate event. He knew that someone on the staff would cover and guarantee that the production continued as normal. In the end the LD healed, and the show continued without him. Remember there will always be more shows, but you only have one life.

Staying Healthy

Every bookstore in the world has volumes of books on how you can remain healthy, but the following are some sure-fire tips that can help when working in our industry.

Wash your hands often – You spend all day touching cables, gear, and a lighting console, which are often handled by others as well. Hand washing or hand sanitizers are imperative, especially before eating anything.

Eat well – Your body can only take a limited amount of cold pizza and burgers at three am. Fruits and vegetables will help replenish important vitamins.

Get some rest – Sleep rejuvenates our body and mind. Skip going to the bar one night and instead go to bed early. Try to achieve eight hours of sleep.

Vitamins are your best friend – Many people take vitamin C to keep their immune system strong and vitamin B12 for stamina. Daily multi-vitamins are helpful too.

Drink lots of water – Our bodies are made of mostly water, so drink as much as you can to keep fresh. Health authorities often recommend eight eight-ounce glasses per day.

Get some fresh air – Sitting in a venue for fifteen hours a day can be both physically and mentally draining. Periodically go outside and sit or walk for 30 minutes to soak up some sun and fresh air.

See a doctor – When you start to feel bad, go to the doctor, or ask the production office to call for an on-site medic. It is much better to have a professional diagnosis and treatment than to continue to suffer.

Limit alcohol and drugs – Some recreational usage is acceptable for many (remain within the law), but extreme use could lead to health problems, lack of sleep, and addiction.

The Show WILL Go On

One key element to remember is that in most cases, the production is about the act or event, not the lights. If you have to go to the hospital, the show *will* find a way to continue. With some simple preventative measures, you can usually avoid these situations, but you should also be prepared for when you are sick. When our bodies are not well we usually do not perform at our best and the production will suffer. Whether a concert, theatrical event, house of worship, or corporate presentation, all productions *can* continue without any specialized lighting. There is no show worth sacrificing your health and life over.

29 Time to Sleep

Working in any portion of the entertainment field, particularly lighting, one will find periods of time when they do not get enough sleep. Lighting professionals are especially susceptible to sleep deprivation as the working duties can take many hours, and often need to schedule around other production needs. Sleep deprivation can be extremely dangerous as it slows down cognitive functions and impairs most tasks. Lighting staff will inevitably be placed in situations requiring less than ideal sleep, and it is good to be prepared for these conditions.

Good Sleep

Experts agree that adults should get between seven and nine hours of sleep a night. This rest ensures maximum physical and mental acuity while also promoting health and wellness. While I generally maintain this sleep goal when at home, I cannot think of any production period where I was able to sleep this much on a regular basis.

Unfortunately, many lighting processes need to happen when the stage is free of other departments. This generally means working late at night after rehearsals, set builds, photo calls, or other needs. Additional workers need light to work and do not appreciate lighting changes while they are trying to see on the stage. Some lighting work can happen away from the stage through the use of graphical visualization. Lighting visualizers have helped by moving the programming into the virtual world. However, there is no substitute for the real rig, and thus much of the work must occur with the actual fixtures on the stage.

During some productions, you may find it best to work as a "vampire"; working during the night and sleeping during the day. Keep in mind that it can be tough to swap your schedule from the norm and then swap it back once rehearsals start. These transitions often cause sleeping difficulties. Anytime you are sleeping during the day, you need to ensure the proper environment. Close all curtains, keep the room cool, and stay away from television, phones, computers, and other screens. Put your phone on do-not-disturb and strive to get as many solid hours of sleep as possible. Proper rest should be a top priority in your life.

Bad Sleep

When you get little to no sleep, your brain begins to get foggy and will not process actions or the world around you very well. You will find it difficult to speak complete sentences, and you may even forget simple tasks during your regular routines. Mistakes will become more common, and your understanding and reasoning skills will suffer. Working long hours without sleep may appear as if you are accomplishing more, but in reality you are introducing more potential errors and problems into your workflow.

Many productions simply do not allow enough time for a proper sleep. For instance, I was working on a corporate event that had a six am lobby call time. We then traveled to the venue, rehearsed some segments, and started the show at ten am. The show finished at eight pm; then we had more rehearsals until midnight. Once this was complete, I needed two more hours of programming to prepare for the headlining band coming the next night. So I returned to my hotel at two-thirty in the morning, only to have another six am call time. A similar cycle repeated for four days! As the days continued, I certainly was suffering from lack of sleep and exhaustion. Plus it was not just me, the entire production crew was affected by these crazy required hours. Not only were we exhausted; we were all beginning to speak differently and make minor mistakes. Going to catering was like visiting a strange zombie land of catatonic people!

Schedules like the one described seem to be more and more common as productions get bigger and more complex. In many cases in my career I have gone multiple continuous days with little to no sleep. Concert touring is built around a similar cycle with full days followed by short sleeping time on a bus. In any of these situations, it can be nearly impossible to achieve the recommended eight hours of sleep on a regular basis.

Sleep deprivation may not only be a result of the actual working hours. Many shows combine a travel and workday for efficiency; however, this rarely provides for rested employees. If you have to travel for six hours and then immediately work another ten, you will be at the limits as to what you can achieve. Furthermore, many of us steal away sleeping hours on the road to go out for dinner, drinks, sightseeing, or other activities. While it may be entertaining, you will pay the price for your night out when the six am call time comes!

I'll Sleep When I Die

When working on a production with limited allotted time for sleep, it is best if you can find some time for a quick nap. With a touring show this is easiest because the bus with your bed is generally not too far away. However, many other productions simply are not setup for napping. Personally, some of the places I have slept include on the floor between the seats in an arena, in my programming chair, on a roadcase, and in the production office. Studies show that even ten minutes of rest help rejuvenate the brain and body.

If you are fortunate enough to sleep on a plane, this is another area to catch a few Zs. I find that when I am tired I sleep like a baby from the moment the plane starts rolling until we land. However, I know others who suffer the entire flight wishing they could sleep just a bit. Sleeping on a plane is a good use of the time, but trust me, it is never equivalent to a nice bed (unless you are in premier first class).

Fighting the Deprivation

Lack of sleep is inevitable for most lighting personnel. No matter how you plan for the deficiencies, sleep deprivation will take a toll on both your body and your productivity. There are many stimulants that can be used during times of limited rest to help keep you awake and alert. Coffee tends to be the preferred choice among many. The caffeine in this readily available drink will provide a more alert state of mind and keep your motor running for hours. Tea, soda, and energy drinks each work in a similar manner by delivering caffeine or comparable substances to our brains to help stave off the effects of sleep deprivation. Be sure to stop the intake of these drinks a few hours before the end of your workday. Otherwise, you may find yourself lying in bed unable to sleep.

As you continue working without sleep, your body will crave sugary and starchy foods such as candy, chips, and other unhealthy items. These items will help your body survive in its sleep deprived state, but can have health consequences later on. Weight gain and increased cholesterol levels are to be expected with longer bouts of less than optimal sleep. Try to stave off the sugary snacks and instead go for healthy energy boosts such as bananas, apples, and nuts.

There are also energy supplements and drugs which enable people to remain awake for long hours. These are extremely dangerous (and often illegal) as they lead to impaired thinking, poor judgment, and dangerous health risks. Although they may be common in some circles, illegal drugs should never be consumed during working hours as an aid to combat sleep deprivation.

So Very Tired

Sleep is a very important human function; just as important as eating or breathing. Do your best to not allow productions to over-work you to the point that you become sleep deprived. Although long hours are common in our industry, it is up to each individual to take care of him or herself and achieve quality rest. Try to nap when you can, eat healthy, and only make sparing use of caffeine and other energy aids. If you must go a night or two with limited sleep, get to bed as soon as possible and make it a priority to sleep for eight or more hours. Remember, you only perform at your best for a production when you are properly rested.

30 When It Stops Being Fun

Working in the entertainment lighting field provides happiness and joy to many on diverse levels. However, there are some unfortunate situations that can emotionally drain you to the point of wanting to quit a gig or lighting all together. Don't let these rare occurrences chase you off, though; instead, use them as opportunities to learn for the future. However, some state of affairs could force you to speak up or quit a show due to unforeseen situations.

If your overall experience and mindset are in good shape, then you will survive to continue along with your career. It is important to identify a downward spiral as soon as possible, then you can make informed decisions to respond accordingly. After all, you want to continue to work on productions that fuel your passion for lighting.

Going Down Fast

No single event or person has ever caused me to change my attitude towards a gig. When this occurs it is the result of many factors all coming together against the greater good of the production or directly against me. Usually lack of sleep and improper nutrition play a role as well. When a coworker yells or is rude, remember that he or she is likely stressed and may not actually mean what they are saying. Take the time to calmly respond to the rant and solve the overriding problem together.

In other cases, there might be personality traits or attitudes that simply do not align with yours. It may be best to keep your head down and complete your job with as little conflict with the opposing person(s) as possible. Thankfully these are rare instances that usually work themselves out with proper communication.

The Emotional Drain

A few years back, I was involved in a production that had many factors come together to knock me down emotionally. The production was a success from the audience point of view, and everyone involved was happy in the end. No one was fired and the show went on without a hitch. However, getting to the last day was definitely an uphill climb. In the midst of the turmoil around and within me, I wrote some notes about how I was feeling. It was an unusual feeling for me, and I truly was unsure how to deal with it in the moment. Here is a portion of my diary entry that summarizes the state I experienced:

> Right now I feel tired, frustrated, and I feel like I have lost all will to care about anything to do with my work here. I really don't like how I feel in this moment. I really don't want to be involved anymore. I now believe I am working for the paycheck and not for the passion inside me. That sucks. I feel beaten to a pulp and that my input, experience, and advice have been ignored at every turn. I don't have to be correct . . . I just want the best outcome for the show.

The passion for a great show that usually lives inside of me had been alive and kicking throughout the pre-production period. I worked hard to ensure a top-notch production while collaborating with all of the production staff. Problems kept arising from every angle, and anything I did or said was challenged due to politics and budget. Throughout it all, I continued programming and operating the lighting, as well as assisting many other production positions that were failing.

The very passion that always burns within me had been nearly extinguished when I wrote that passage above. It saddens me to think that I was driven to that low a level. After the show ended, I took some time to contemplate the occurrences and debrief with others where possible. I also decided to not work with certain members of that production in the future.

Speak Up

Imagine you are busy working on a show or in a venue and you find yourself going down a similar terrible path. The first thing you should do is to calmly speak up with other members of the production staff. It is easy to sit around and complain, but this never solves anything. The best thing to do is to call a production meeting and calmly list the problems at hand. Explain that the production is in trouble and that you have reached the end of your rope. Ask for help in a calm manner, and I bet you will be surprised at the results.

Too often people let their emotions drive them and they begin yelling and making others mad. Calmness is key here. In my previous example, we managed to solve many problems in a meeting I called where I explained peacefully exactly what was going on. Other members of the staff were then willing and able to help resolve the complications that had existed for days. My relaxed presentation of the problems helped set the tone for the meeting and encouraged others to refrain from raising their voices.

Sensory Overload

In my early twenties, I learned to skydive and jumped out of an airplane at 11,000 feet. The moment I pushed off the wing and became free, my mind went nuts. My instructor said this is known as "sensory overload" because of the flood of new inputs from the senses to the brain that are occurring simultaneously. I always liked that term.

In early 2001, I had the opportunity to program the tour lighting for the Crystal Method with lighting designer Lawrence Upton. We were given two weeks to pre-program the show in a small studio in Los Angeles. The complex electronic rhythms of their music provided many opportunities for interesting and detailed lighting. We were programming at the rate of about one hour of programming for each minute of a song. Their production was all about the light show, so all the lights were frequently pointing out towards the audience. We were tasked with overloading the senses of the audience during the show. This was the early days of EDM music and production, so you can imagine that the lighting included many intense movements and powerful strobe effects.

After about a week of working back to back fourteen-hour days at this pace, we came in for another day of work. Suddenly we discovered that Lawrence and I were bickering with each other like an old married couple. We both were completely blocked from thinking in a creative manner, and had barely managed to write a single cue in the first few hours. It became apparent to both of us that we had been in our own form of sensory overload. After spending every moment of every day staring into bright flashing lights, listening to loud music, and avoiding sunlight and the real world, we had reached our pinnacle.

So we decided that no further programming would occur that day. We turned off the rig and headed out to the beach. We spent the entire day soaking up the sun and sights, and then had a relaxing meal. The next day we arrived back at the studio refreshed and ready to go again. We were able to program the rest of the show with a renewed sense of creativity, and we both learned the value of giving our senses a break. Try to avoid too much sensory overload without giving your brain time to recover.

Watch Out for Murphy

When frustration, sleep deprivation, emotional disruption, and other negative influences strike, be on the lookout for Murphy. When we are in these states, it can be difficult to respond appropriately to the unknown situations that occur according to "Murphy's Law." An LD who might be adept at dealing with a power outage at FOH could overreact if he is already disappointed with other problems on the show. In a normal emotional state, he would be able to recover accordingly with little impact on the production. However, when he is already drained and beaten, he is likely to make further mistakes as he tries to recover from this new problem.

Be on the lookout for the unexpected and try to respond instead of reacting. Once you have a negative outlook about a production, then suddenly everything you see tends to favor the negative. Remain as optimistic as possible, and don't let several small problems appear as one big depression zone.

Be Aware

The real key to remaining happy in the entertainment lighting field is to be aware of your own emotional state. There are millions of self-help books on this very subject because humans find this process difficult. Every moment of life cannot be happy and perfect, thus we must learn to analyze situations so that we can restore as much joy as possible into our lives.

In addition to learning lighting skills, take the time to work on yourself for the betterment of all around you. Books, podcasts, seminars, therapists, and other resources are all available to help you grow. Then if you practice becoming aware of your own internal feelings, your newfound insight will establish a better relationship with all for your next production.

Once you identify how the production environment is impacting you personally, then you can take action to correct it. Many LDs, programmers, and other lighting staff have made the decision to walk away from a production in order to preserve their own emotional state. Once they realized that the conditions were harmful to their industry passion and internal emotional state, they chose to heal themselves by removing themselves from the negative environment. Their careers then blossomed even further as their internal passion burned stronger than ever. Be on the lookout for joy-draining circumstances, and take action as soon as possible to restore your passion.

Part VI

The Travel

When you are working in the entertainment lighting business, you can guarantee that at some point you will be asked to travel for your work. Even venue employees are sometimes tasked with attending a tradeshow or working an event at another property. More often, freelancers will travel to where the gigs happen, either through flights or on a tour bus (or even a combination of both).

Traveling in the entertainment lighting industry runs the gamut from the cheap to the luxurious. Some will start out staying in low cost motels and sharing driving duties with the band members in a small van. You might have to share your hotel room with others and eat fast food. However, the average traveling person in our field will stay in decent hotels, travel on buses and/or commercial flights, and devour fresh catering. Overall, the traveling is very reasonable and comfortable.

On the other end of the spectrum, certain designers, lighting directors, and crew will earn the opportunity to upgrade their travel experience. Some will travel on chartered flights, ride in limousines, and stay in five-star hotels. In extreme cases, a few will travel with the artists on private jets, attend exclusive meals and parties, and generally follow the celebrities as part of their traveling party or entourage.

The majority find traveling in the entertainment lighting field very enjoyable. In nearly all cases, the production (or the company that employed you) should pay all your travel expenses. Never should you have to shell out your own money for your job. Be sure to get reimbursed for taxis, ride shares, baggage fees, parking, and other common travel expenses. The following pages detail the very basics you need to know to travel like a pro.

31 Travel Is Work

First and foremost, you need to understand that the very act of traveling is labor, and should always be thought of as such. Spending your time waiting in lines, going through security, sitting on planes, riding to/from the airport, and the entire travel minutiae occupy a major portion of your valuable time.

Traveling Value

Always negotiate payment for your travel days. You certainly cannot work other gigs while you are getting from A to B, so the current gig should compensate you accordingly. While you may not always earn your full day rate for traveling, at least half is totally acceptable. If you are paid a weekly rate, then ensure that the travel days fall within the paid week, or that you are paid for the entire week in which you travel. The paid week should not start after you arrive on-site.

You might decide to do some work while on the plane or at the airport. Drawing plots, programming consoles, editing content, creating patch sheets, and other computer related tasks are easy to accomplish during your commute. Remember, however, that the actual travel is your task at the moment. Depending on your contract, you might have been able to charge for those hours if you had done the work at your desk before departing. Once you are getting the travel rate, you likely will not get reimbursed for additional work you perform during the voyage. Of course, if you use the travel time wisely by working on a different production, then perhaps you can "double-dip" by charging one client for your travel time and another for your office time.

Time Is Money

Some customers may ask you to travel and work on the same day. They will book you to fly early in the morning and then expect you to come to the venue at noon for load-in. Or perhaps they ask you to run the lighting for the corporate event and fly out a few hours after the show ends. These timesaving measures will task you immensely, thus you should ask for appropriate compensation.

For instance, if during one day you travel for four hours and then work for eight, it should not count as a single twelve-hour day. Instead, you should charge for your travel rate plus a day rate. Let the client simply save on hotel costs, but don't allow yourself to work for less money than you deserve for your time and effort. Reasonable sensibility should always be applied here. If your client asks you to fly all day and then attend a quick production meeting, it is probably not worth asking for additional money to cover both.

When working for a company, as opposed to being a freelancer, you will need to abide by the travel policy of the business. As a salaried employee they will see your travel time as part of your job. I still advocate viewing the journey as work and suggest not multitasking while traveling. You should instead manage your time accordingly so that you can relax and focus on the travel versus overworking yourself.

32 Earn the Perks

The travel industry offers many rewards, incentives, and upgrades, and you should grab all that is entitled to you. It is strongly recommended to sign up for as many travel service reward programs as possible. Airline miles, hotel points, rental car plans, and other loyalty programs reward you with perks and benefits that you may not otherwise be allowed to enjoy. I have been very fortunate that, over the past twenty years, I have not had to pay for flights or hotels during my vacations. Instead, I have made use of points and rewards to cover these costs.

Airline Miles

Airline frequent flyer programs are the most lucrative for any traveler. Beyond earning points for flights, they also present benefits such as preferred seating, no-baggage fees, upgrades, minimal lines, free drinks, and early boarding. Additionally, if your flight is delayed or rebooking is needed, having higher status will certainly aid you in these rough situations. Remaining loyal to one airline will earn you the highest status; therefore, you should try to fly the same airline (or its partners) as much as possible.

This may be easy if you are booking your own travel, but often business travel is booked through a production's travel coordinator. You can always ask for bookings on your preferred airline, and many will accommodate as long as the fare difference is not too great. Even if you are booked on another airline than your preferred one, be sure to sign up for their frequent flyer program to guarantee you are earning points on that system too. When flying, the miles are generally applied to the passenger and not the booking party, so make certain that you are earning all that is available to you. Then remember to use these miles the next time you need to book personal travel.

Hotel Points

Hotel points are useful as they can be redeemed for free stays. As with airlines, it helps if you can remain loyal to one brand of hotel. This way you will gain the most status and opportunities for upgrades, bonus points, and other perks. Hotel points are often more difficult to earn because they frequently apply to the payment holder's account. If you are traveling on tour, or if the production staff books the rooms, then it is likely you will not earn points for your hotel stay. This is because the room rate permits only the booking agent to earn the points on the group package.

However, if you are working for a company that allows you to book hotels on a corporate or personal card, then the points will likely be yours for the taking. Similar to airline miles, be sure to sign up for every brand of hotel in which you happen to stay a night. Keep track of your loyalty status through a smart phone app, and sign up for extra bonus point programs when they are offered.

Also remember that any purchases made at the hotel can usually earn points, whether you booked the room or not. If you buy meals, sundries, or hire laundry services your spending will usually be eligible to receive points based on these charges. Always provide your membership number to make sure you get credit towards your purchases.

I once was able to unlock expired airline miles because I purchased ice cream at a hotel gift shop while on tour! Thankfully, I had earned three points because I provided my hotel membership card when I bought the ice cream. These points then enabled me to reactivate airline miles due to a partnership between the hotel and airline. You never know when a simple purchase will enable further perks and benefits.

More Benefits

Rental cars, shared ride services, airport parking, restaurants, and other vendors that you utilize during your travels usually have reward programs. The more you can make use of these (especially when spending per diem or expensed dollars), the more you can take advantage of free or discounted items in your personal life.

In many cases, points and miles can also be redeemed for many different services and products. Sites such as points.com allow you to exchange your loyalty points for cash, services, gift cards, and products. I know of one lighting technician who traded in his airline miles for a new water heater in his home!

It's So Easy

Sign up for all the various programs you can find to maximize your reward possibilities. Then store all the membership cards in a convenient location and install the various apps on your smart phone or tablet. Check your status often to confirm you are receiving points and sign up for additional bonus programs when offered. Within no time, you will amass enough points and status to gain further perks and joy from your travels.

33 Air Travel

Flying has become the most common method of transportation around the world for business. It is also often touted as the safest way to get from point to point. However, many people find the entire experience extremely stressful and aggravating. If you know how to properly manage the time, flying can easily become another simple task in your business and personal life. The following outlines some tips to help you have the best flying experience possible.

Pre-Flight Planning

Always select your seat and double check that it is correct at the time of booking. Whether you are booking the flight or if it is booked for you, the first thing you should do when receiving your itinerary is to check that the assigned seat is one you actually want to sit in. Usually you can log into the airline website or app to see a seating chart. Depending on your status, you may be able to select better seat for no fee (otherwise, additional costs will be shown).

Sometimes a travel coordinator will book a flight without selecting seats, and this will default you into a middle seat in the back of the plane. You are the one who will sit in the seat, so take the time to select a seat that you would be happy to plop yourself into. If your preferred seat is not available, keep checking back (especially a few days before the flight) as new seats may open up. On a recent flight, I managed to change to a much better seat the day before my flight departed.

When packing, decide if you are going to carry on your luggage or check a bag. Checking a bag usually will cost additional dollars, but this fee is often reimbursable with the production or company that employs you. Frequent flyer status, credit cards, and other perks may allow you to receive free bag checking as well. The choice to carry on or check may not even be an option if you are hauling tools or other specialty equipment not allowed in the passenger compartment of the airplane. Otherwise the choice really is an individual preference.

Personally, I prefer to check my bag unless I am on a direct flight or in a hurry upon landing. I find that competing for overhead bin space can be stressful, and I also don't like to roll my suitcase all over the airport. Regardless of what you choose, your suitcase should have your contact information on and in it as well as some personal identifiers on the outside. If you can easily distinguish your bag from others, it will reduce confusion for yourself and other passengers.

If you are a frequent traveler, it makes the most sense to have a separate set of toiletries and other common items instead of having to pack and unpack them for every trip. You can just refill the small containers as needed between trips and keep your suitcase ready to go. Learn to efficiently fold your clothing and only take the essentials on your journeys. Multiple shoes and large coats add bulk along with weight, and should be avoided unless absolutely required. Never pack your essential phone and computer charging cables in a checked bag. Always carry these in your backpack, as you never know when you might need to plug something in.

At the Airport

Arriving at the airport one and a half to two hours before your departure time should provide adequate time to check bags, navigate security, and trek to your gate. Of course, common sense should be applied as some airports or departure times require additional time. For instance, early Monday mornings at all airports are generally busier than Wednesday evenings. If you are checking bags, it is normally fastest to use the kiosk system to check in instead of waiting for an agent. In many cases, the kiosks are also faster than the priority or first-class lines too.

After checking your bag, you will need to head to security. There are generally three or four different methods of access (and a few others that are out of site). If you are simply a general passenger then you will be in the longest line. Some frequent flyer and credit card perks will enable use of the priority line, while many also sign up for a pre-check system such as TSA pre-check (www.tsa.gov/precheck).

With a pre-qualification system the security line will move faster as passengers have been previously screened (and paid a fee) to reduce the scrutiny required for airport entrance. Because everyone in these lines is not removing his or her shoes, electronics, food, and other items it tends to move faster. There are also higher priced services that allow you to cut in line at the checkpoints. The biggest VIPs have secret security areas with little to no wait, but those are normally out of the average person's reality.

Once through security, you will be free to move about in the terminal. It is best to get close to your gate area and confirm that your plane is on time. Food and drinks at the airport are expensive, but you should always get some water. Because you cannot bring liquids through security, you will need to either purchase the water or bring a container to fill. Many airports have bottle filling stations, which are easy and free to use. Others simply have standard water fountains, but these often dispense foul tasting water. Additionally, you can usually ask a coffee shop or other vendor to fill your water bottle for no charge using their filtered water.

Unless you have an ailing body, I strongly advise standing and walking prior to your flight. Personally, I find walking in airports a very relaxing and enjoyable activity. This is also why I often check my bag, as I don't want to drag it around as I walk for an hour. With my backpack on and ear buds in my ears, I set about to walk the entire expanse of the airport before boarding my flight. About ten minutes before boarding, I will stand in the gate area waiting to board the plane. By the time I am in my seat, I am usually ready to sit down for several hours. I have never understood why some people choose to sit for an hour at the gate before boarding a multi-hour flight only to sit longer.

On the Plane

Prior to boarding the plane, you should prepare yourself with essential items for your flight. These items will aid in your comfort, entertainment, safety, and prepare you for the unexpected. A small backpack or bag with these items can be placed under the seat in front of you for easy access. Plus the majority of these items will also be useful at your work site too.

Always bring water and a snack on every flight, no matter the length. I was recently on a two-hour flight that included a five-hour delay sitting on the plane before we took off. Although in this case the airline did pass out water and snacks, often they do not. You should maintain control of your own food and beverage options and not rely on the airline. Plus, if you happen to have a long flight with a meal service, there are good chances that the food might not be to your liking.

The flight attendants will warn you to fasten your seatbelt anytime you are in your seat. Trust me, you want to do this! It does not have to be tight, but significant turbulence is always possible, and you don't want to be thrown about. Make it a point to stand up for at least five minutes every three hours, and always be cordial to your fellow passengers and flight crew. If something goes wrong on your flight, don't panic. Look around at the flight crew to determine the severity of the disruption and follow all their instructions. Remember that flying is continually proven to be the safest mode of travel. I have been in several emergency landings, and thankfully none resulted in major problems.

Flying is a great time to catch up on personal needs such as reading, podcasts, music, and movies. You may be inclined to work during the flight, but often this is difficult due to the limited space and prying eyes of other passengers. Some people are able to sleep on planes (thankfully I am one of them), and flying is a great time to catch some sleep. Remember though, that unless you are in a first-class lie-flat seat, the sleep you get will

Table 33.1 On Plane Essentials

Item	*Purpose*
Noise blocking or canceling headphones or ear buds	Drowns out background noise and allows clear listening to entertainment of your choice.
Ear plugs	Nice to quiet your space when you are trying to sleep or read.
Water	Always bring your own water. Purchase bottles at the airport or use a refillable container.
Snacks	Nuts, protein bars, and other munchies help reduce hunger.
Entertainment devices	Smart phone, tablet, laptop, electronic reader, and other items help pass the time.
Reading materials	Books and magazines can provide entertainment and/or learning.
Power cords, plugs, and adaptors	Bring what you need to plug in your devices at the airport and on the plane. Always anticipate needing a charge.
Pen	When traveling internationally, you may be required to fill out landing documents.
Pillow	Some like to bring a travel pillow for improved comfort.
Warm clothing	Planes are often cold. Have a light jacket or beanie cap in your bag in the event of cold air.
Hand sanitizer	Always a good idea to keep healthy. Apply before eating anything on the plane.
Cleansing Wipes	Wiping down the area around your seat helps to reduce the spread of bacteria and viruses.
Medicine	Bring essential medications with you and also have a standby of pain relievers, anti-acid, nausea, and other common remedies.
Flash drives	If you are responsible for maintaining data such as plots or show files, have backups on flash drives with you.
Batteries	Some devices may require a battery change, or a backup battery for additional energy when plugs are not available.
Bag tag receipts	Always keep the tag that the airline provides when you check your bag. This will be essential if your luggage is missing.

(Continued)

Table 33.1 (Cont.)

Item	*Purpose*
Computer	Never check your laptop or tablet; always bring these with you on the plane.
Socks	If your shoes do not require socks, you might want a pair when going through security or if relaxing on a long flight.
Sleeping aids	Certain drugs, herbs, and foods will help you get some rest.
Toiletries	It is best to bring some basics with you in case your luggage goes missing, or if you simply want to freshen up in the lavatory.

not be as restful and rejuvenating as you might expect. Always make the most of your time on the plane so that you have a great experience during your journey.

Arrivals

When your plane arrives at your destination airport, take the time to wash your hands and use the restroom before collecting your luggage or finding your ride. If you are required to pass through immigration, consider that lines are often long and prepare accordingly. Then once you are in the baggage claim area, stand away from the carousel and watch for your luggage. Keep an eye to ensure another passenger does not grab your bag by mistake. Once you see your luggage, step forward, and grab it. This may seem like common sense information, but take a look around at how many people crowd the carousel where the bags drop. It can be quite chaotic, and this is why I recommend standing back from the mayhem.

Once you collect your luggage, locate the area for your transportation and head out of the airport. You might have to meet up with other production staff to share a ride, or look for a runner or chauffer holding a sign with your name. Keep any receipts from your journey and be sure to get reimbursed for all your travel expenses.

If your bag has gone missing, don't worry, as it likely is coming in on another flight. Head to the luggage office and provide them your tag receipts (you kept these, right?). They will ask for a description of the bag and any other important information. Often they can tell you when the luggage is arriving, and sometimes it may be within the hour. In some instances, it may have even arrived before you on an earlier flight.

Otherwise provide them your hotel/destination information and they will deliver your luggage as soon as possible (usually later that same evening). If you are on tour though, it can be very difficult as your destination may change every day. In this case, you might find it best to provide the address of a hotel or venue two days ahead to ensure that the bags catch up to you. When your luggage is missing, ask the airline for an amenities kit that will provide you with basic toiletries and perhaps a t-shirt too. Rest assured that very rarely does a properly labeled bag disappear forever.

34 Tour Buses

Many in the entertainment lighting business will earn a living traveling the world on a tour bus. While the buses may look different in various parts of the world, the basics of this mode of transportation remain the same. Your first time on a tour bus will very much be an outsider experience. You will find seasoned roadies who are extremely comfortable with the lifestyle, and they may not share all the aspects of this transportation method right away. Tour bus living is different than other types of travel and it is important to understand the uniqueness.

Bunk Life

Living on a tour bus is comparable to sharing a small apartment with a group of people that you probably would never choose to live with. The total number of people on a bus will vary from tour to tour and can be anywhere from six to twenty. Respect for each other's personal space, food, habits, and routines are essential. Although much of the time on a bus is spent sleeping, there are often long rides during which you will be "living" on the bus. This is when it is crucial to learn to share with your bus mates, while also trying to live your own life.

Typically, there are three main locations on a tour bus: front lounge, bunks, and back lounge. Some parts of the world utilize two-story buses, which further help people to spread out. Your luggage will be stored under the bus in the bays and there may or may not be room on the bus to stash your backpack and other essentials. Sometimes an empty bunk becomes the "junk bunk" where riders can store medium-sized bags.

A small kitchen area provides meals during long travels, or even when you want to skip catering. Most tours will supply the bus with essential food and drinks as well as provide a bus-stocking list. You can make requests for specific bus food and the production team will fill the bus accordingly. If you request something you don't want to share, be sure to label it appropriately. Additionally, most productions provide a "warm" meal after the show ends. Pizza, sandwiches, spaghetti, and other meals from local restaurants are often placed on the buses for the staff to enjoy after completing their work.

At the start of a tour, bunk selections and assignments will occur. Sometimes this is an organized process, and at other times is a free-for-all. In either case, the general rule of seniority normally applies as the most seasoned staff gets first pick. Your bunk will become your personal space for the duration of the tour. You may choose to store a few items here; but not too much as space will be limited and you will want to have room to sleep. The driver will periodically clean your sheets and pillowcase as well as the rest of the bus. For optimal safety, always sleep with your feet towards the front of the bus.

About Number Two

In terms of the restroom, the number one rule on tour buses has to do with number two. A long-standing decree of touring is to only put liquid waste in the bus toilet. If you feel the urge to do other business, you will need to ask the driver to find a rest stop for you. Although bus technology provides non-smelling systems and easy disposal, this rule is still in effect. Maybe at some point in the future the regulation will be overturned, but until then don't do number two on a bus.

Shared Space

It is everyone's responsibility on the bus to ensure the environment is safe and friendly. You should always lock the door and luggage bays behind you when leaving the bus; secure it as you would your own house. Before inviting friends, fans, family, and others onto the bus for a visit, confirm that your bus mates are okay with strangers entering the shared home. Introduce your visitors, and don't allow them to roam free on the bus, eat the food, or use the restroom. If a person unknown to you enters the bus, ask them whom they are with, and escort them off the bus if they cannot name a bus mate or they lack the proper production credentials.

Generally, people from the same department will travel together on the same bus. This is because the lighting staff generally need to arrive and depart around the same time, while the backline crew may finish load-out earlier and can get rolling sooner. Sometimes you might share the bus with members of the band, which may sound exciting but generally this is not the case. When you are on the band bus, you don't have a chance to remove yourself from your work. You may wish to relax after a show and not continue discussing their thoughts and requests. Furthermore, the artists often have their own needs and desires that may trump yours, simply because of their status.

Some tours will have buses that cater to specific segments of crew. For instance, there may be a smoking bus or a sober bus. These designations help those with specific habits maintain their lifestyle while not causing anguish for others. In other situations, smoking may be allowed only in the back lounge. Some people even partake in drug use on buses with a feeling of immunity. Tour buses are not sanctuaries and are open to searches by law enforcement and border patrols. Furthermore, underage drinking, illegal drugs, and other offenses are usually not tolerated on buses. These actions can result in termination and/or prosecution.

Living on a tour bus can be an enjoyable experience as long as everyone on the bus gets along and has respect for each other. If you are having difficulties with specific people on your bus, you can ask the production manager for reassignment. Quite often though, great friendships and camaraderie are formed on buses. Even with great bus companionships, you will find hotel days filled with a new appreciation for the comforts of privacy and aloneness. Tour bus travel is not for everyone, but it can also become a wonderful way of life that many enjoy throughout their career.

35 Hotel Living

Staying in hotels is a regular occurrence for many in our industry. Personally, I have averaged over 100 nights a year for way too many years of my life. While staying in a hotel may seem straightforward, there are some things with which you need to be aware. First, the production will generally book your hotel and will have already paid for the room. You will be responsible for any further incidental charges to the room. Be sure to always pay your hotel bill, and settle with the front desk before your departure lobby call.

When production books your hotel, have them provide you with the booking confirmation number so you can check in with ease. This way if there is no record under your name, the front desk can look up your stay with the confirmation. When traveling on tour the production manager will usually check in everyone at one time and then distribute the room keys upon arrival. You may receive your key as you enter the hotel, or perhaps they will be provided on your bus. Either way, make sure you collect your key as soon as it becomes available. Remember, the production manager wants to get to his or her room quickly too.

In the Room

Fairly often our lighting work requires working through the night and sleeping during the day. When your schedule flips from the norm, you need to prepare accordingly. First, bring some gaff tape or binder clips to your room to help blackout the curtains. Hotels are notorious for curtains that do not completely seal out sunlight. Another trick is to use the hanger with clips the hotel provides to seal your curtains shut (see Figure 35.1).

Always place the "do not disturb" sign on your door and alert the front desk to the fact that you are sleeping during the day. This way they will not call to check if you want the room cleaned or disrupt you in other ways. Soft earplugs are helpful too as hotels often perform maintenance during the day, which could further hinder your sleeping plans.

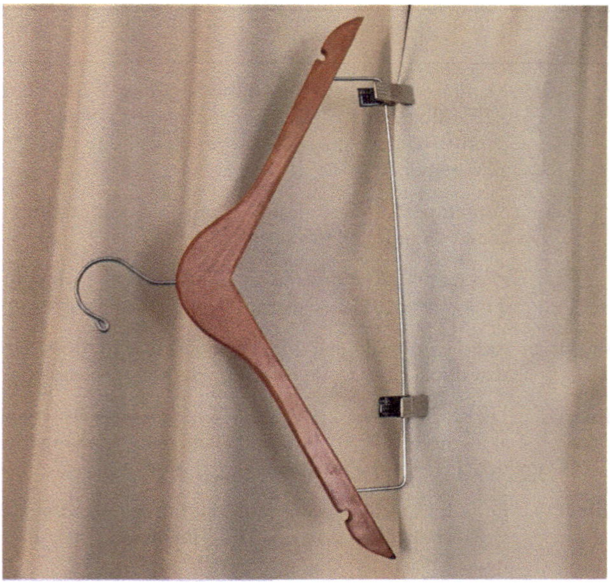

Figure 35.1 Hangers with Clips are Great for Closing Curtains

Hotel rooms may seem like a secure location for your belongings, but always use vigilance as with any other location. Keep the door locked and the do not disturb sign on the door (even when you are in the room). If you need additional towels or a cleaning, just call the front desk. I find that, in most cases, I am only in a hotel a few days and do not need staff entering or cleaning my room. Make use of the hotel safe and hide your laptop and other valuable when you are not in the room. Don't forget to collect all your belongings and double-check the room prior to checking out.

Sleep Well

Remember that the main goal of your hotel stay is to get a good night's sleep. You can always ask to move to a different room or floor if you cannot get comfortable in your assigned room. When you check in, ask for a room with a king size bed, as this will likely result in a larger, more comfortable room. Actual room type upgrades generally will cost you money, unless you have loyalty status or points to use. Don't hesitate to ask for the things you need to ensure good rest. For instance, if you have allergies you can often ask for hypoallergenic linens or rooms sanitized with natural cleaners.

Table 35.1 Helpful Tips for Hotel Stays

Item	*Tip*
TV remote	Many find that these can be full of germs. Wipe it down with a cleansing cloth or place it in a plastic bag. Some hotels allow TV control via your smart phone so you can skip touching the remote all together.
Earplugs	Sometimes hotels are noisy. Disposable, soft earplugs will help you achieve a quiet night's sleep.
Hair dryer	Always point the dryer away from your head when first activating. This way any dust or hair inside will blow away from you.
Lock doors and windows	Keep your doors and windows locked at all times. Double-check that doors close securely when you leave the room.
Report problems	Let the front desk know about any problems you might experience at the hotel. Often, they will compensate you with loyalty points or vouchers for the restaurant or gift shop.
Hide valuables	Do not leave important items out in the open. Place them in the safe or behind chairs or curtains. Remember to claim your stashed items before checking out!
Override thermostats	Many websites explain how to overrule hotel thermostats to achieve better temperatures and disable motion detectors.
Do not disturb sign	Use this sign at all times for security and to keep unwanted hotel staff from your room.
Cancel housekeeping	Many hotels provide bonus points or extra perks if you call and cancel the daily housekeeping.
Spare key	Some rooms require a key placed in a master switch to enable the power in the room. Ask for a spare key to insert in this slot so that the power (and AC) remains on at all times.
Extra items	Forgot something, need another small bottle of lotion, or want a late-night snack? Ask for what you need, as hotels are great at providing hospitality for their guests.
Late checkout	If you know you are leaving after ten am, ask for a late checkout at no charge. If you must leave your room, consider checking your bags at the bell desk until you depart the hotel.
Ask the concierge	Hotel staff often know about the local area and can offer discounts or advice about restaurants, walking areas, laundromats, and other sites.
Use the gym	If you like to exercise, make use of the provided hotel amenities.
Bring a long cable	Travel with a six-foot or longer charging cable for your phone so that you can place it where you prefer.
In-room clock	Always check that the time is correct on the clock in the room. Also ensure the alarm is turned off.

(*Continued*)

Table 35.1 (Cont.)

Item	*Tip*
Use your own alarm	Use your own travel clock or smart phone as a wake-up alarm Never trust the hotel clock or wake-up calls.
Wear sandals or slippers	Hotel rooms are notorious for spreading foot fungus. Bring your own slippers for the room, or stand on towels when not wearing your shoes.

As mentioned previously, always sign up for hotel loyalty programs so that you can earn perks, points, and upgrades. With many hotel chains, simply being a member provides you free Internet, breakfast options, bottled water, and other perks. Utilize the hotel app on your phone for additional features such as room selection, digital key, and hotel information. It is also helpful to setup a personal profile that details the type of room and location that you prefer. For instance, you can designate if you like high or low floors and whether you wish to be near or far from the elevators.

Hotel stays are daunting for some, but routine for many others. Make the most of every stay by envisioning the room as a sanctuary for your sleep. With a goal of achieving the best rest possible, you can enjoy the hotel and take care of yourself. Both the production and your body will be grateful that you achieved some sleep.

Part VII

The Interviews

A career in any industry has many twists and turns through various positions, opportunities, and experiences. People have their own goals and insights, which drive them to very different paths throughout their employment journey.

I selected eight key people in the entertainment lighting field to interview and profile about their specific lighting career path. Each has a tremendously interesting story to share about how he or she has achieved success. They all also continue to progress forward with only slight thoughts about when retirement will enter their path. There are millions of stories in our industry, and these eight are but a few examples to inspire you within your own career.

36 Cory FitzGerald

Production and Lighting Designer, Creative Producer and Creative Director, Senior Partner at Silent House Productions

Starting at the age of four, Cory FitzGerald was introduced to Broadway shows by his parents, and he quickly formed a fascination with the theater. With his eyes originally set on being a Broadway lighting designer, he has had a varied career that now sees him moving beyond lighting to producing and directing major productions and tours. Looking back through his career, FitzGerald says:

> I realize that the combination of desire, skills learned, luck, relationships, and just being in the right place at the right time has brought me to this point. I have always been driven to work hard and to become better at my craft.

Figure 36.1 Cory FitzGerald

The passion for stage lighting started when Cory was twelve, after several years of attending a summer theater camp in Los Angeles. He became friends with the program's lighting designer, James Schipper. James offered him the opportunity to work on the crew, running follow spots, hanging lights, and ultimately learning consoles. On the day of his sixteenth birthday, FitzGerald started working at Angstrom Stage Lighting in LA and expanded his knowledge of everything lighting related. FitzGerald remembers, "I was still in high school, but doing all kinds of shows and working all around town in various theaters whenever I could."

After graduation, FitzGerald went to NYU (Tisch) in New York to study lighting design. Here he learned about the detailed world of theater and focused on smaller shows downtown and off-Broadway. His drive to learn more inspired him to sign up for the (first and only) Automated Lighting Academy held in Los Angeles during his first summer of college.

> This was my first experience with a Wholehog 2 console, and it opened me up to the future of moving light programming. The Academy was hosted by High End Systems and featured the top LDs and programmers as instructors.

Upon his return to NYC, he leveraged his new relationship with High End Systems and began working as an intern in their New York City office. During this time he was able to form relationships with many of New York's top lighting designers. This led to a full-time job in product support for consoles and (the then emerging) media servers. Upon graduation from NYU, Cory worked for High End Systems in New York, and then London, focusing on control software and support for programmers and designers around the world.

> I knew and was grateful that I was in a fortunate position where I got to both learn and teach new and developing products. Plus I was able to interact with designers, programmers, vendors, developers, and artists on what worked, what didn't, and what could be better about all sorts of products; then follow through on actually helping make substantial changes.

FitzGerald eventually decided to leave High End Systems and pursue a freelance career as a lighting programmer. Armed with the connections and relationships he had fostered, he was sought after by a variety of designers and directors for some remarkable opportunities.

> While having the skills to use and understand the consoles and fixtures, I realized that I had a lot to learn about working with and for others on an artistic level. I quickly started to develop new skills to help

> streamline my work flows, multitask, and quickly process what was going on around me. Fortunately, a string of amazing opportunities came my way, and I continued to branch out into new and larger, more complex gigs. Thinking I would still end up in New York theater, I continued to push for more work on theater shows, eventually getting to program a few Broadway shows and national tours.

Then, in 2006, a defining moment happened in FitzGerald's career during the programming of Madonna's "Confessions" tour. He was the second programmer on the show working with lighting designer LeRoy Bennett and his team. For FitzGerald, this was the first time he really unlocked the potential to program lighting to music in an incredibly precise and creative way.

> I had been a musical person all my life, playing various instruments and appreciating a wide variety of music, but in a matter of weeks I found myself being able to create looks and match lighting to sound as I had never been able to before. I quickly figured out it was the path I'd been waiting for and it didn't take me long to be pulled into the touring world full-time.

The next five years of his life were constant programming of touring and television shows for a handful of different designers. FitzGerald eventually felt the urge to take on the responsibility of designing shows himself. He wanted to step beyond his comfort zone and push himself creatively. After reaching out to his contacts, indicating he was looking for a lighting design opportunity, he realized that it would be difficult to take a step back, but necessary to make a new step forward. However, he had confidence in himself, which he believed helped him with the transition.

> I had never "designed" a music show on my own and knew few would want to hire me without design credits; even though I had programmed hundreds of large scale shows. I started by reaching out to friends or clients looking for a lighting designer for smaller or emerging artists that would be willing to take a chance on someone like me.

After a designing a series of small shows, FitzGerald experienced another lucky twist of fate when he received a call to design for an emerging artist named Bruno Mars. He interviewed with the manager and got the gig. As lighting designer, programmer, and operator, he worked with Bruno to help create his first national tour, and the live events and appearances that followed.

> Working with Bruno pushed me past what I had experienced up till then. He was a true musician, using no timecode, and a full band with

Figure 36.2 Cory FitzGerald's Career Path

> amazing artists who could riff and change arrangements in real time, and I was told to "keep up." The experience of pushing myself to do things I simply "had to do" completely changed my perspective once again so that I could meet, and hope to exceed, the artist's, the audience's, and my own expectations.

While Mars's career exploded into more tours, TV shows, appearances, and music videos, FitzGerald was there learning new and exciting practical and life skills. Over time, he had moved beyond the programmer role and had become a respected lighting designer.

> It was like going to school all over again, but by staying focused on the constantly changing project in front of me, I really didn't notice till it was all behind me.

When Bruno took breaks from touring and shows, FitzGerald started working for other artists, and quickly had a full schedule of productions for many top artists including Beyoncé and Jay Z, Justin Bieber, Demi Lovato, Gwen Stefani, Skrillex, Janet Jackson, Jennifer Lopez, and Kendrick Lamar, to name a few. FitzGerald has also worked on a number of movies, globally televised award and talent contest shows, as well as Super Bowl half time performances.

FitzGerald has added more skills to his repertoire and now works as a creative director and creative producer. He enjoys the opportunity to work closely with the artist in creating the concept of the show from his or her vision. He understands the music and helps create set lists, mood boards, renders, and drafting to turn ideas into reality.

> Loving music and creating shows comes from passion, but the experience required to do it well and efficiently takes time and practice like anything else. Understanding the business, the budgets, the collaboration and all the moving pieces is so critical to a successful outcome. The best way to do it is to try, learn from your mistakes and successes, and keep trying to out-do yourself whenever possible.

Relying on a mantra instilled in him long ago, FitzGerald shares the following career advice that has guided him through his professional and personal life. "When there is no solution, there is no problem," and this simple, yet slightly sarcastic statement helps him through tough times.

> It does make a situation seem more manageable to me when facing a no-win scenario, where simply the best we can do with what we have is the way forward. Nothing in life seems to end up as perfectly as you may have planned it, but being able to rely on past experience, think on your feet, multitask quickly, and roll with changes that are

> out of your control puts you in a far better position than attempting to move mountains when it's simply not going to happen.

His work has been his life, and FitzGerald sees the two as integrally intertwined. As he explains, "It's a one hundred percent lifestyle that you quickly get initiated into. You need to learn to meld work responsibilities with life's events. When you love every moment of your work, it's never just a job."

There is much more ahead for FitzGerald and he reiterates the importance to keep learning at every avenue. He is always looking for new information or concepts and offers:

> Always keep your eyes open, ideas are all around you, and light is everything you can see, literally. Listen to music, study products, go to shows, look at art and nature, and try to see how everything relates to your craft.

37 George Masek

Strategic Marketing Manager

George Masek's career has ranged from lighting technician to product specialist, and he is now a well-known and respected figure in the automated lighting industry. His personal path in the field is as unique and interesting as any.

> I started with absolutely zero contacts in the business and no way at all to get in, or even learn about the job I wanted. From those beginnings, I was able to build a career that has been rewarding and that I've been proud of. I had some great luck, but I had some potentially tough

Figure 37.1 George Masek

> breaks too. The key was truly that I was very clear of my initial desired destination and I NEVER took "no" for an answer. If I was told "no," I simply found another person to ask the question to, or another way to ask it. As I look back, that is simply the truth on how I made it into the business and then excelled.

From a very young age, Masek discovered a love for lighting and grew a desire to work for the only automated lighting company he knew about: Vari-Lite. But he was far too young to start working professionally. As early as grade school, he built his own lighting system and did shows for free. His first real "gig" was lighting his sixth grade dance. When he was only fifteen years old, Masek snuck backstage at a concert and met the lighting crew chief. Instead of throwing him out, the crew chief offered him a job!

> From there, I said yes to everything that had to do with lighting or even doing shows. I was dresser at a theme park, I hung Christmas lights and decorations at malls, I did pool party lights for kids – absolutely anything. Every gig was a way to meet a new contact, or to learn something else. It was like a giant ladder to be climbed rung by rung. I always said yes and I always got paid for my work – from day one. The jobs grew in stature and I started to meet more and more influential and successful people.

One of those people turned out to be an audio engineer who changed Masek's life forever. This sound engineer had previously worked on a Van Halen tour and worked with Showco, which was a sister company with Vari-Lite at the time. Masek expressed his career goal to work with Vari-Lite and his frustration at being turned down multiple times. One two-minute phone call from this particular audio engineer to his contacts at Showco led to an interview and job at Vari-Lite for Masek.

Standard procedures for new hires at Vari-Lite back then were to attend a ten-day training followed by a placement on a tour as a junior VL technician.

> I apparently showed promise, and after five days of class, it was determined I was "ready." Immediately, I was sent to work the Fleetwood Mac tour that came to Dallas. I was told I would have a one-day tryout and if things went well, I would be joining the tour.
>
> As I watched the progression of the day, from load-in, to show, to load-out, I realized quickly that I could cut a lot of time from the whole day. At the end of the night, I told the console operator that I hardly knew anything about fixing our lights, but I was certain I could reduce 45 minutes from the load-out with a few changes.
>
> He agreed, told VL to put me on the tour, and off I went. I am incredibly proud of that first experience as it paved the way for the rest of

> my career. I was completely unprepared, utterly terrified, and had to rely on my wits, luck, and the kindness of total strangers. What I DID have as the absolute desire to make it no matter what.

Masek continued to tour with Vari-Lite for the next two years when he realized he had a career choice. He could continue on advancing as a programmer, lighting director, lighting designer, and show designer. Or he could stop touring and move ahead on the corporate side of Vari-Lite as the company and the industry grew.

Starting on the production line in Manufacturing, Masek learned all that he could about the lights he had been touring with. He became an expert on the assemblies in the fixtures and could repair them as quick as anyone. He found himself learning directly from the Vari-Lite product engineers and became a sponge for new information.

He was soon promoted to Tech Support where he was able to interact with tour technicians, programmers, lighting directors, and even designers. Before long he moved up to a position managing the Tech Support department. "I didn't realize it, but I was in the process of building lifelong relationships and friendships with the touring and production communities," explains Masek.

As the industry grew, so too did the need for automated lighting in markets other than touring. Masek next took a position as a support manager for their new leasing business. Many productions including Broadway, television, Las Vegas installations, theme parks, and cruise ships jumped at the opportunity to add automated lighting to their shows.

> This job also began my transition over to the "Sales" side of the business as we were primarily in the business of selling (leasing) to these customers. My boss began to groom me as a salesperson and I was given the role of account manager. I started supporting tours for Lord of the Dance, Jimmy Buffett, and more. Once again, it was a baptism by fire – I was a technical guy, and suddenly, I was dealing with the financial side of things – rentals, personnel, trucking costs, sub-rentals.

After some time in sales, Masek realized that he had stepped to far from the light and needed to get back to lighting.

> I liked the increased pay and the challenges were interesting to me, but I wanted to be more involved in actual lighting – or at the least creating the lights and being part of that process. Mostly, I decided that I wanted to help make cool tools for the amazing artists and technicians that I knew. I seemed to have the ability to talk to them, and then translate what they were telling me to the product engineers.

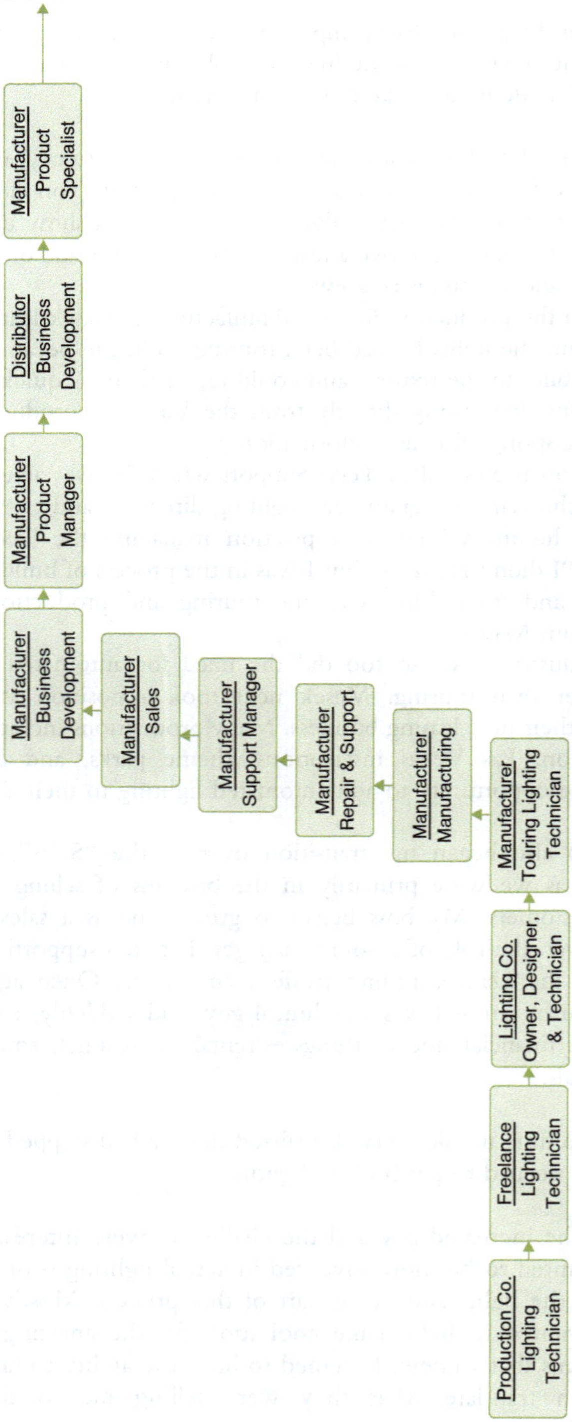

Figure 37.2 George Masek's Career Path

He transitioned into a new role at Vari-Lite as an overseer of their top clients. He became the point person for key accounts regarding fixture issues, support, and customer needs and wants.

> I saw myself primarily as the facilitator to get all of these really talented people to work together to fix whatever issue we faced on a particular show. I enjoyed this role tremendously and suddenly realized how beneficial all those classes unrelated to stage lighting that I had taken in high school and college were to me. All of that knowledge helped me be very effective at my job.

Then the unexpected happened; Vari-Lite was sold to new owners and his job was changed to product manager. All his experience, knowledge, and contacts would be put to the test as he helped to develop many iconic products for the company including the VL3500 Spot and Wash and VLX. With more time, Masek eventually left Vari-Lite and began working with a distributor of Clay Paky fixtures. Again, his product manager skills were valuable as he assisted in the development of products such as the Mythos and Scenius.

Masek has most recently transitioned to working directly for Clay Paky as a Senior Product Specialist where he continues to interface with designers, programmers, and technicians.

> In the end, the role combines production, sales, engineering, and product development – a little bit of all of the things that I've done in my career. It is interesting to me how all of the experiences and relationships over the years have helped to prepare me for this role. And while it isn't anything like what I set out to do, it is exactly what I am very fulfilled and content doing in the industry.

Thirty-eight years after forming a love for lighting and a desire to work for an automated lighting manufacturer, Masek is very happy with his career and thankful to all that have helped him along the way. He wraps up with this advice, "You have one shot at this life – dream as big as you can possibly imagine and don't let go of that dream – ever."

38 John Viesta

Lighting Designer and Broadway Associate Designer

Diversification keeps John Viesta happy and interested in his career. Since the late 1990s he has worked on countless Broadway shows for a "Who's Who" of lighting designers as well as built an impressive resume of his own designs. "If you have your toe in a few different ponds, it is easy to change directions and try different things," explains Viesta.

Early on, he knew that he wanted to be a working professional in the lighting field and have a family. While in college at Purchase College he had desires to work on Broadway as a lighting designer, and started forming a plan to make that happen.

Figure 38.1 John Viesta

> I have continuously built goals for myself. In my early twenties I was always careful that if I hit a goal, that I had another right behind it. I am the type of person that needs to work towards something. I feel very fortunate to have been given the opportunities to achieve where I am today.

Purchase College once served as a tryout location for many Broadway shows, so he was exposed to Broadway level shows as early as his sophomore year. In fact, an opportunity presented itself that summer to work as an intern with lighting designer Natasha Katz as a moving lights tracker on *The Scarlet Pimpernel.*

> This was the first of many sink or swim experiences, when, as a nineteen-year-old kid, I was thrown into a full, legit Broadway production. I did not even know what a moving light tracker was supposed to do. Natasha was very kind and willing to teach me while letting me learn and gain responsibilities.

Throughout his time in college, Viesta was presented with several more opportunities to work with Natasha and other Broadway designers. He negotiated with the school to allow him to finish his curriculum while also working on shows away from campus. During this time he also started working with Paul Gallo, which continued for nearly ten years.

Right out of school, Viesta started working with Theatre Projects Consultants on many different installations. Working on the design of theater spaces helped him to understand how shows interrelate with the venue they are held within. He again negotiated the ability to continue to work on other shows at the same time as his duties at Theatre Projects Consultants.

While he was climbing the ladder and growing his career, he found some difficulties within his own family. They did not understand or support his career in the arts. They wanted him to learn and explore, but were concerned about his financial viability. "In my early twenties, there was a lot of pressure from my family to not go into the business," Viesta recalls. "Understandably, it is extremely competitive, and many people fail."

However, Viesta did not fail. After a few years, he decided to leave Theatre Projects Consultants to focus on his lighting design aspirations. He made himself "mutually exclusive" to many designers at the same time. Natasha Katz, Paul Gallo, Ken Posner, and Peter Kaczorowski each hired Viesta and taught him along the way.

> I got so much experience seeing everyone's different styles, which allowed me to distill my own style. Each person I assisted worked extraordinarily differently from the others. They are all so talented and successful, but how they work and their creativity are all uniquely different. I find it an interesting thing to observe.

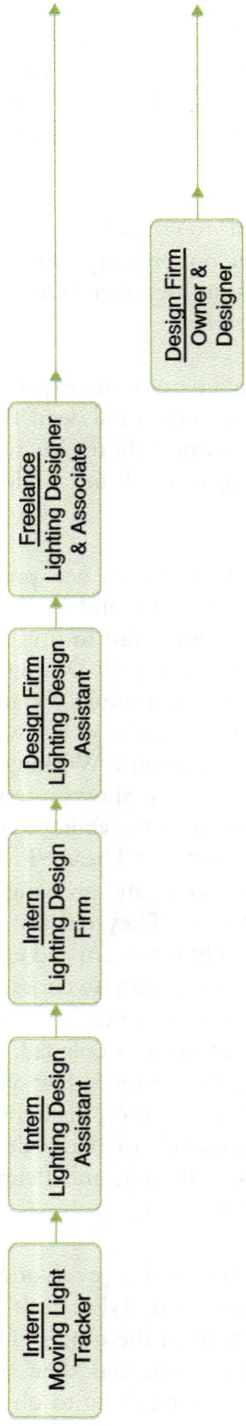

Figure 38.2 John Viesta's Career Path

While moving up in the Broadway field from an assistant to an associate designer, he also continued to design for off-Broadway shows and other productions. However, many of these smaller productions were not fiscally beneficial. Viesta found he had to carefully choose which projects to peruse, often weighing creative opportunities against financial viability.

> Early on I wanted to design on Broadway, but I quickly recognized that there are only so many seats at that table. There is a stable of very talented designers that work in that circle and it is a hard nucleus to penetrate. That's when I discovered that diversity can help me and that I should try other crafts.

He began branching out with his own design projects outside of theater and found a new trajectory for his career. Viesta now finds himself designing multiple stages for IMG Fashion Week in NYC, working in television for networks such as ESPN, ABC, Bravo, NBC, and even designing web-based video productions for obscure cliental. Plus he continues to be busy as a top associate lighting designer for many Broadway shows.

Viesta has found his happy place with a work/life balance that allows him and his family to live the life they want to live. Their financial situation provides what he needs for his daughter, and he is very satisfied with the work he is doing and projects on his plate. In fact, his parents are now proud, even if they don't fully understand the actual work he is doing.

As with life, Viesta's career has not been as straightforward as it may seem on these pages. In 2010, he was assisting Peter Kaczorowski on a show when he was granted a rare Saturday day off. Before enjoying his day away from the theater, he created a backup of all his paperwork and even joked, "Just in case I get hit by a bus." He did not know how prophetic that statement would be.

> Saturday morning I went for a bike ride when an inattentive driver struck me on my bike. It was a head on collision; my face went into the windshield and I flipped over the car. I was seriously injured and in the hospital for a few weeks. The impact should have killed me. It crushed my skull and all my facial bones were broken. They had to totally reconstruct my face.

Even after he was released from a long hospital stay, his team of physicians would not allow him to live alone. He had to live with his mother for a few weeks and was unable to work. After four months of recovery he was finally able to return to work. However, he had no income during this time and his life savings were cleared out through his recovery. His medical bills amounted to over $360,000. Luckily, he had insurance through his union, and the driver's insurance covered most of the bills too. But living costs, co-pays, and other expenses still added up greater than what he had available.

He then discovered the Behind the Scenes charity and applied for a grant. The organization went beyond just providing financial support for a month and a half. They also provided much needed emotional support. Viesta then had to rebuild his life and career again after this terrible instance. Many of the designers he worked with were very happy to work with him again and offered him plenty of work.

With a new lease of life, Viesta jumped back into his career and has continued without looking back. He has no idea what his profession could look like in ten years, but is excited to see what happens. He is saving through his union's pension and 401(k) plans, while also living a budget oriented life.

In reflecting on a career in the entertainment lighting industry, Viesta says, "It is critical that you want this path. You have to want it more than anything. It has to be in you and be exploding out of you." Viesta has certainly demonstrated that lighting is essential to both his personal and working life.

39 Robert Roth

Industry Pioneer & Senior Account Executive

Robert Roth's career has followed along with the evolution of the modern lighting industry. His path has taken him on an amazing journey that helped set the standard for many ways that we use light in productions today. Looking back, he says, "Minimalism is great and can be lots of fun, but nothing beats spectacle." This statement can be applied to most of his life, and his entire career has certainly been quite the spectacle.

Figure 39.1 Robert Roth

In the late 1960s, Roth was thirteen and inspired by the oil projection shows he saw at various rock shows. He knew he could do better, so he borrowed some projectors from his junior high school and started doing shows. Always the entrepreneur, Roth named his first business the "Aurora Light Show."

A few years later, he went to a shop to purchase new professional speakers, when he saw a pile of lighting gear on the floor. There were SCR dimmer packs, a small control board, along with Fresnels and ellipsoidals.

> I walked in to buy the studio monitors and left with the lighting gear; which magically was the same price as the studio monitors. Right then, I decided that I was going to be in the theatrical lighting business.

He promptly called a friend who owned a booking agent and asked, "Which bands need a lighting guy and can afford to do that?" Roth was connected with an up-and-coming band that led to three years of touring with the act. During this time Roth graduated high school and continued working with the band instead of going to college.

As the band's run came to an end, Roth was offered an opportunity that would further change his life. An owner of an audio company approached him and said that since he had a lighting company they should work together. At the same time, Roth's father offered him a small stake of his college savings to start his own business. At the age of twenty years old, Roth formed R.A. Roth and Associates in his parents' basement.

Within ten months, he had a warehouse and was building gear from scratch. This was a time when there was no organized lighting industry, or even lighting designer positions. His company provided gear, lighting services, and invented the much of the industry infrastructure and procedures still used today.

> We had multiple lighting systems; we were building the gear from scratch. We were building our own road racks, using multi-pin connectors, and creating square housing PAR fixtures. My dad was an engineer and we even started designing trussing. He got us a structural engineer and made sure everything was certified. And off we went!

The company transformed into R.A. Roth Inc. and they cut their teeth on many of the top country and R&B acts of the day including Willie Nelson, Rick James, Triumph, and more. Roth invented new systems and procedures to ensure his designs and company stood out among the competition. For example, when most tours of the day were using large grid arrays of PARs, he began creating vertical looks and pods. Creating scenic elements out of lighting intrigued him, and his design concept continued for many years. In his warehouse, the innovations continued as they pioneered the concept of load and patch testing in their shop, and developed new methods for packing trucks.

Over time, the position of lighting designer was becoming understood, and Roth found more work than he could handle. He began working with Jim Chapman and Steve Kaye on shows outside of his lighting equipment provider company. Even with the calls coming in, they wanted more and set their eyes on an upcoming Michael Jackson tour (that had not even been announced yet).

> Michael Jackson just released his *Thriller* album and everyone said there was not going to be a tour. I didn't believe it. I said, "We need a project like that to take our careers where I want to take them." We worked in secret for seventeen months to figure out how to do that. Lo and behold we got hired!

The stadium tour took them to the next level as designers, and Roth and Chapman, along with business manager Kaye, designed many big tours throughout the 1980s including Madonna, Ozzy Osborne, Whitesnake, as well as televised productions such as Live Aid. As moving lights developed, they made use of early Vari-Lites and also expanded into set and production design. Because of Steve Kaye's pushing the firm into early involvement with personal computing, they were also the first designers to make use of AutoCad in concert design starting in 1984.

Eventually the designers went their separate ways, and Roth focused on his lighting equipment and production company. They continued to provide equipment for many different shows and tours for several years. Then, after a series of challenges, he made the choice to sell his company to PRG in 1997.

Roth embarked on a new phase in his career as he was employed at PRG as a client manager. His relationships and contacts, combined with his real-world experience, made him a natural for this role. After eleven years, he felt it was time to depart from PRG and began to look for new options.

He was introduced to Huntly Christie in late 2007 and began working with Christie Lites in 2008. His role continues there as an account rep managing clients and shows. He has complete control and responsibility for the projects that he is involved with. Roth does not design lighting today, but instead provides input on the look and logistics when asked about an LD's vision. He also guides designers to excel within their budget.

> The ability to reinvent myself and become successful over and over, and my incredible resilience in this industry, has been the most surprising aspect of my career path. Today I have great career satisfaction; how could I be so lucky at this point?

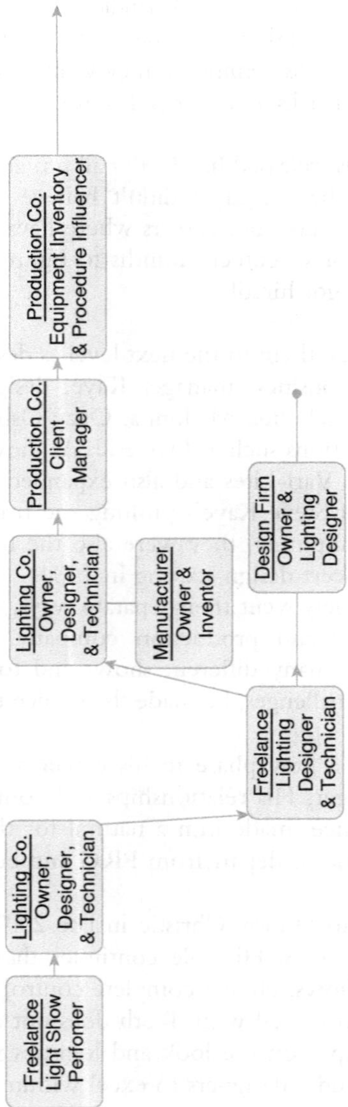

Figure 39.2 Robert Roth's Career Path

Roth points to relationships as a major strength throughout his career. He recommends staying connected with peers, clients, and everyone you come across.

> Business is made in person. It is far too easy for people to forget that in today's technologically connected society, business is made in personal human interactions. A thousand emails can be replaced by a hundred phone calls, or three face-to-face meetings. If you want to forget that, do so at your own risk!

Roth is not looking to slow down or retire anytime soon. Instead, he is holding his 75th birthday as a marker for his next review point in his life. He does say that as he ages he has less tolerance for abusive situations and avoids these as much as possible. As with all of his career, he is neither bored nor unhappy and looks to keep his life that way as long as possible.

40 Ron Schilling

Lighting Crew Chief

In May 1994, Ron Schilling was hired by Upstaging for ten days to help prepare the Eagles' "When Hell Freezes Over" tour. A week prior he had filled out an application on the advice of a friend. Once the trucks were loaded at the end of the prep, he asked the shop manager, "What do I do now?" He was told, "Keep punching the time clock until we tell you to stop." Now, over 26 years later, Schilling is considered one of the top lighting crew chiefs at Upstaging and the touring industry.

Figure 40.1 Ron Schilling

> During those ten days I knew in the back of my mind that I was going to get a full-time job at Upstaging. I had been doing lighting since the second grade. There had always been something about it that I wanted to do. I never said no to anything, including during those ten days.

Schilling discovered lighting in the second grade when he noticed the strip lights and knife switches over the choir stage. He became involved in stage crews throughout junior high and high school. Then instead of going to college, he started with Upstaging, finding the road a much better classroom.

> It's really hard to teach this stuff in college. There is better knowledge on the road because technology moves so fast. Plus life and how to get along on the road is not often taught in schools.
>
> Lighting can be fickle in a lot of ways when you get to really big shows. There is a lot of responsibility to quickly come up with answers as people are counting on you for many things. You must learn how much preparation you need to do to come up with those answers for any given scenario. If the rig goes dark two minutes before the show, you need to know how to resolve the problem quickly.

Schilling did not instantly become a crew chief, of course, and instead had to climb his way through the ranks. Once he started touring as a lighting technician, he took opportunities to work with opening acts, help other departments, and never said no. He has always been vocal about "throwing his name in the hat" for other positions he finds interesting. Most recently he has taken on lighting director duties for a Guns N' Roses tour.

A life of touring is not for everyone and Schilling is okay with the compromises he has made in his life. Touring over 300 days a year has resulted in missing life events, not getting married, taken a toll on relationships, and caused him to miss holidays, weekends, and birthdays. But there are also unique and exciting experiences that come with traveling the globe.

> Touring is full of compromises and sacrifices that you have to make, but also opportunities that most people will never ever get. If you do this enough, you will eventually have a day off in a town you have never seen. Getting to know the band members and learning how nice most of them are is also a treat.

In 1997, Schilling had a moment that opened his eyes to the "reality" of our jobs and responsibilities. While on tour as a technician, he found his crew chief dead after he had fallen out of a house catwalk. Fortunately,

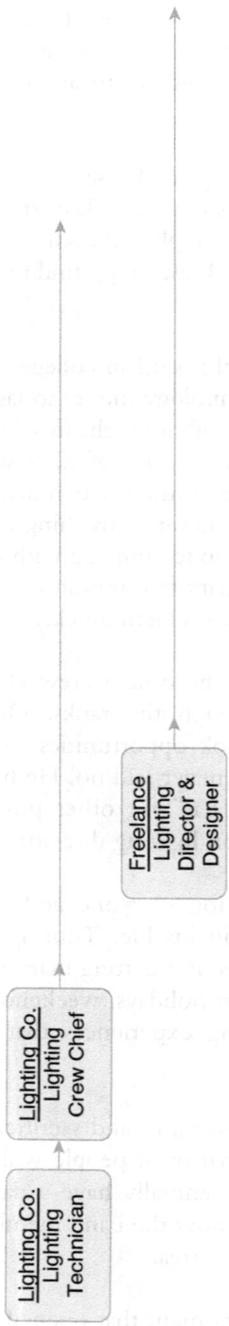

Figure 40.2 Ron Schilling's Career Path

safety standards and practices have improved tremendously since then, but the dangers are still prevalent.

> We fly thousands and thousands of pounds of equipment above people's heads and one lapse in judgment or little mistake can be quite costly. You have to know yourself, your gear, and your responsibilities. Just because you have a harness on does not mean you should be in the air.

Some of the highlights of Schillings career include a world tour with AC/DC in 2010 where everything just clicked, touring with Def Leppard (he's a big fan), and the first time he ran a lighting console for a Guns N' Roses show in 2016 at Dodger Stadium.

During his touring life, he has been preparing for retirement by saving and not overspending. He owns multiple rental homes and has contributed to a 401(k) plan through Upstaging for most of his life. He calculates that he should be financially set to retire when he is 55. He is also now looking at transitioning to a lighting director/designer role as opposed to performing all the manual labor as a crew chief. He might just decide to continue on for many years in one of these new roles.

A good mantra that Schilling wishes to share is, "Don't be afraid to make mistakes, but do what you are doing for a reason." He has certainly created a great career by following this credo.

41 Sooner Routhier

Lighting and Show Designer

Surprisingly, I wanted to be a marine biologist. I had aspirations to attend school in the Virgin Islands to study coral reefs. I think it's always been about color for me. I loved the vibrant colors of fish and corals in the reefs. Plus, my mom was a seamstress and worked at a fabric store when I was really young. I used to love poring through all the colored fabric. Now I know that I find those same joys in colored light.

Figure 41.1 Sooner Routhier

This comment from Sooner Routhier clarifies much of what has led to the successes she has had in her lighting career, which began in 1998.

Growing up in a small town in Northern Vermont, Routhier would travel at least two hours and cross into Canada just to attend a concert. At the time, she and her friends had no concept of career paths in the lighting and production industry. It was not until attending college in Vermont that she was first exposed to the field. During their spring concert, the school would bring in audio and lighting gear as well as professional staff to help support the show. It was at this time that she met lighting designer Brian Clarke, who let her "tag along" on other shows to begin her learning the craft of lighting for music. "It was definitely a right-place at the right-time moment for me," says Routhier.

After a few months of on-the-job training, Routhier was hired part-time as a lighting tech and operator for Rainbow Production Services. Four years later, she transitioned to a freelance position and quickly was sent out by Port Lighting to tour with the Irish Tenors. Routhier explains, "They sent me on my first tour as the lighting tech. I never looked back! Once I had the touring bug, I was hooked."

After the Irish Tenors tour, more opportunities presented themselves to tour with more bands and in increasing positions. Throughout her career, Routhier has found networking with others to be an essential skill.

> While on tour, I continued to expand my network. I would offer help when it went beyond my work duties. I would always take the opportunity to meet with fellow roadies on days off. I took full advantage of meeting production managers and other designers on festivals.
>
> Eventually, my social skills paid off and I met my business partner, Robert Long. He hired me to be the substitute LD for Audioslave's tour in 2005. We continued to work together on multiple tours. Then in 2009, we decided to form a design company. As of this point, we have been working to build our company for over ten years!

Now Routhier is an accomplished lighting designer and moving on to full production design. Some of her favorite productions she has designed include the Muse Simulation Theory World Tour (lighting design), Game of Thrones Concert Experience, and Imagine Dragons Evolve tour.

> My role changes from project to project. At times, I'm a show designer. I design all aspects of the show with the artist, create production scripting, determine lighting fixtures and cuing, sketch and create scenic elements, assist with video content and storyboards, even help with pyro and automation cues. Other times I work only as a lighting designer alongside a production designer or creative director. I draft lighting plots, spec fixtures, and program the cues for the show.

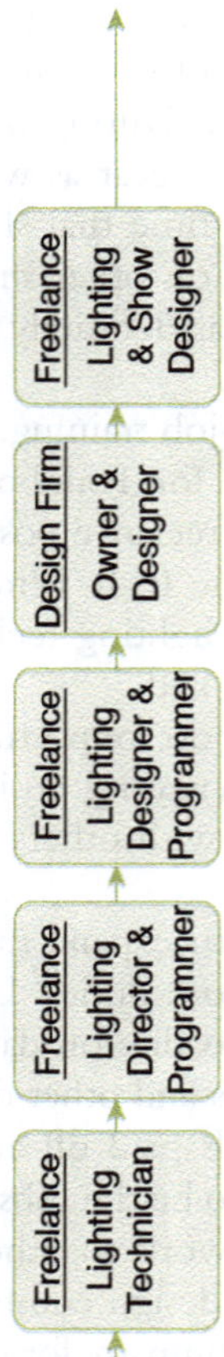

Figure 41.2 Sooner Routhier's Career Path

> My favorite thing is to create an entire production as a whole. I love telling a story with production elements. You can really take an audience on a journey by placing specific moments in specific time slots throughout a show. Depending on those moments, you can carry an audience through different emotions paired with music and/or action.

Not every moment along her path has been ideal. She recounts a time that was big low for her in the moment, but also a big growth moment afterwards. During her first year working professionally, she was asked to substitute on a show for a staff member who was unable to attend. Quickly she found that she was in over her head as the rig was too big for her to handle on her own. "I remember crying in a corner because I didn't know how to patch the dimmer rack properly. The lighting designer with the touring act was pretty upset that he couldn't get the rig working," recalls Routhier.

Eventually she thought through the process with a clear head and got it all sorted out.

> It was one of the most embarrassing moments for me as I really thought I was tougher than that! In the end, I learned more about thinking through troubleshooting properly than I would have under the guidance of someone with more knowledge than myself. Trial by fire – although it sucks – is a REALLY important learning process!

With over 21 years in the business, Routhier has never been happier. She feels that she has finally achieved a strong work/life balance while working with some incredible creative directors and designers. She continues to learn and grow, and appreciates the wealth of knowledge of younger up-and-coming designers and programmers.

> The biggest advice I can give is to never stop learning. I feel that some people stop learning at a certain point in their careers. In a business that is constantly changing with technology, it's important to change with it. You can't stay stagnant in a business that changes as much as it does.
>
> I also think it's important to learn the skills of collaboration. There is so much work out there for everyone. Musical artists collaborate all the time with other musicians to create incredible music. Remember that when someone asks you to work with another designer or creative director it's not an insult. It's an opportunity for each of you to learn and grow creatively!

With no plans to retire anytime soon, Routhier can likely be found working at a FOH or enjoying the Smoky Mountains with her dog and husband. Either way, she will be surrounded by beautiful colors and learning all she can to grow her career and life.

42 Steve Woods

Lighting Designer and Professor

Steve Woods says that he loves happy accidents, and his career path has certainly been filled with such events. When talking about his designs he says, "I love those moments in which something unexpected happens within the design. Something I did not plan. It just happens and it looks great or solves a problem."

In fact, the very reason he started a career in entertainment lighting is because of one of these happy accidents. He was in college studying English Literature and needed a job to help pay for his education. He was presented

Figure 42.1 Steve Woods

with two options: the cafeteria or the theater. After a year working in the theater's scene shop, he felt the pull, and changed his major to theater.

Since that early beginning in 1977 he has amassed quite the resume of design work. Having worked around the world in many genres such as theater, film, television, opera, dance, concert, and architecture, he says, "I have lived life by my own standards and had a blast doing it."

While in college, he was granted a career break when his program allowed him to take fifteen weeks off out of the year to travel and work in the industry. He was hired by the Hartford Ballet as a technical director for their upcoming season. During this time he did three national tours and helped plan the company's international tour too. He not only met many scenic and lighting designers, but this job also introduced him to the Limon Dance Company during the 1982 World's Fair. After 25 years, he is still designing lighting for them.

> Limon taught me not only how to design for dance, but also how to adapt a design when on the road around the world. I learned how to keep the intent of the design with half the equipment in the air. I learned to make fast decisions, work efficiently, and I learned to listen.

His path through the theater world has been rather straightforward moving from summer stock to LORT to NYC. He enjoys designing new works and being part of the process of the development of new plays.

It has not always been easy, though, for Woods. He has seen periods where a genre of work would "dry up" and he would need to seek out new opportunities. He recalls one time after a production that work opportunities were scarce. After several months with no work, he started contacting anyone who might need his skills such as photographers, ad agencies, museums, and clubs. Woods recounts:

> Through my research and desperation, I found two advertising agencies that opened doors both in industrials with Paul Mitchell and Aveda. I had discovered the runway. It was great, a freeing form of lighting design that played to my strengths, dance and concert.

Around his thirtieth birthday, Woods found himself at a crossroads. He was questioning if he really could make a living as a designer. He even considered leaving our industry and attending law school.

> I had to make a decision of how my future would play out and if touring would still be in the picture. Let's face it, touring is a young man's game.

Instead of retreating and becoming a lawyer, Woods returned to school, and earned an M.F.A. from the University of New Orleans.

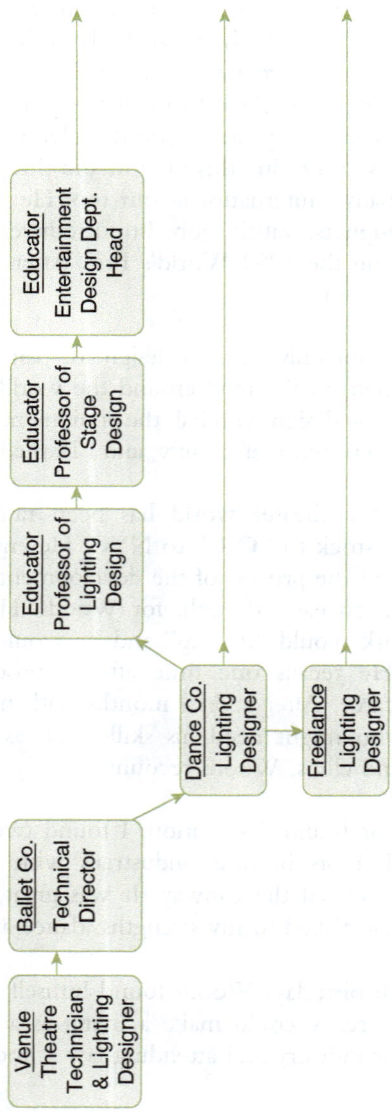

Figure 42.2 Steve Woods' Career Path

> I was even accepted into a Ph.D. program but came to my senses and did not follow through. That moment of decision would ultimately lead to my career in teaching. A career that has given me the financial freedom to pick projects on which I want to work, own a home, and grow a family.
>
> I taught at Louisiana State University from 1988 until 1998, a posting that taught me how to teach. Then, with the arrival of our daughter, we made the move to Dallas, Texas and Southern Methodist University. Which has been one of the best decisions I have ever made. It is a storied program with wonderful colleagues, and class after class of talented and creative design students.

Woods continues as a Professor and Head of Stage Design for SMU in their nationally recognized MFA Stage Design program. He says that all the students he has taught since 1998 enrich his life. In looking back at his career, Woods comments,

> If I were to second guess my life it would be to question, did I take every opportunity that came my way or did I lack the courage to push into uncomfortable areas of growth. Did a fear of failure hold me back at times? But I also think that is normal. It is just part of growing up and learning how to live life.

And his warm spirit shows that he has learned how to live fully. Woods offers the following advice for a fulfilled and happy life:

> Come to work every day believing what you are doing is important. Listen, be curious, and don't forget to be kind. Every day when I shave, I look in the mirror and say, "Today, I will try not to win the jerk award."

Steve Woods hopes to retire later in life to spend time in the mountains with "lots of snow and bears." Until then, he continues to live a very happy life "turning on light bulbs" and getting excited every time the curtain goes up on opening night.

43 Vickie Claiborne

Visuals Designer and Training Specialist

Vickie Claiborne's first lighting design class at the University of Central Florida instantly captured her attention and started her dream of a career in entertainment lighting. She graduated UCF over 30 years ago with a Bachelor of Arts degree in technical theater and a vision of being a touring lighting programmer.

Figure 43.1 Vickie Claiborne

Upon graduation, she started as a board operator for a local Orlando band called Heart Attack Diner, but her life as a programmer really began in 1992 when she moved to Branson, Missouri. Claiborne explains:

> I worked with major country music artists at a showroom called the Grand Palace; that's where I really learned my craft while honing my lighting programming skills, starting with an push-button controller and Intellabeams from High End Systems along with a Celco Gold 90 console.

Before she knew it, Claiborne was offered the opportunity to tour with country artist Kathy Mattea. This required a move to Nashville, Tennessee, but she was finally a touring lighting programmer. Then an unexpected opening presented itself to Claiborne and she jumped at the chance for a new experience.

> In 1994, I got the opportunity to join High End Systems as a Lighting Programmer and Training Instructor; training and programming with products ranging from Status Cue and Cyberlights, to Wholehog 2 and Catalyst.

For nearly a decade, Claiborne was at the forefront of the industry's leading products. Most importantly, she created and defined training principles and procedures that are established throughout our field today. Her legacy can be found in training classes and with instructors across many markets. During the emergence of media servers and digital lighting, she became an expert with the technology and helped shape many of the programming procedures.

Unfortunately, there were some hard times for Claiborne as well. During an economic downturn and company changes in 2004, she became a freelancer once again.

> The years between 2004 and 2008 were full of uncertainty and anxiety for me as I reevaluated my career, my goals, etc. But I never lost sight of what made me happy, and for me that is being creative. It just took me some time to work out what the next stone in my footpath would be, and for me, joining PRG in 2008 was the right move.

In 2008, she moved to Las Vegas and started working at PRG as a product specialist in lighting and digital media. This role allowed her to flourish and learn a multitude of products, including the GrandMA2 lighting console and PRG's Mbox media server. Additionally, her teaching skills were put to use, as she coordinated and oversaw the PRG Las Vegas internship program.

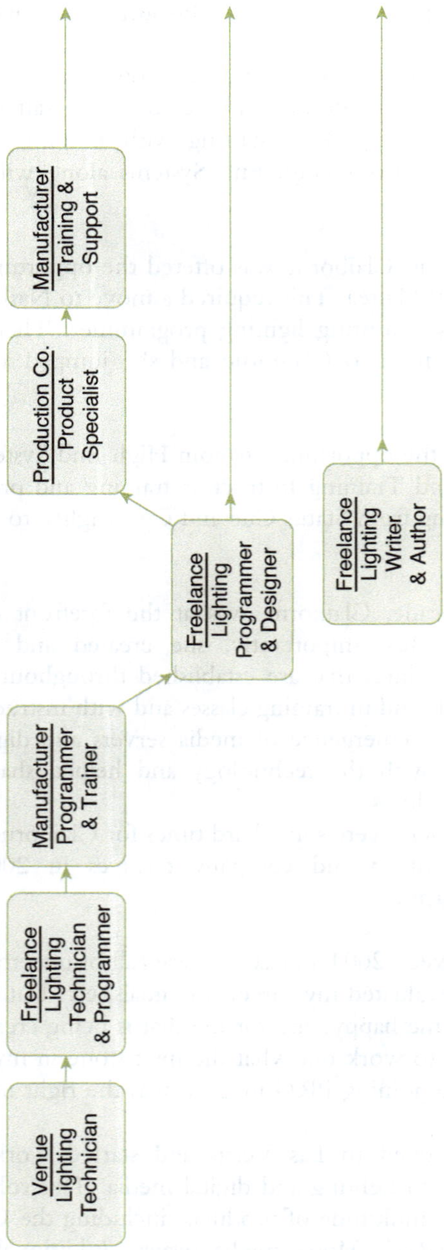

Figure 43.2 Vickie Claiborne's Career Path

Throughout her career Claiborne also continued to program and design productions around the world. Her love of programming and creativity has kept her perusing new opportunities whenever possible. She also has shared her knowledge and experience beyond training sessions by writing a column for PLSN magazine and publishing a book titled *Media Servers for Lighting Programmers*.

Early in 2019, she made yet another career move, by leaving PRG and joining Disguise, where she is employed as their training manager for North America. Once again, she is teaching the industry on a leading product while contributing to the evolution of programming the combination of lighting and video.

Claiborne has always had a focus on doing what she loves and teaching others. She finds it very important to never let your ego be bigger than your talent, and suggests treating others as you want to be treated. She does not regret any moments throughout her career and is grateful for all the opportunities.

> I would have had regrets now if I hadn't taken advantage of the opportunities that presented themselves to me along the way. I tell others now who are just beginning their careers in this business to not worry about the money; instead, focus on doing what you love and the money will come.

Some of her favorite programming moments include the Sydney 2000 Olympic Opening Ceremonies, a sheik's daughter's wedding in Saudi Arabia, and Rock in Rio Vegas and Lisbon. In early 2019, she fulfilled a dream by doing her first show at Madison Square Garden in New York City where she programmed and operated the video for Barbara Streisand via a lighting console.

Claiborne's passion for creativity and teaching will carry on for many years, but she also hopes to reach a point where she will be able to spend more of her personal time on things like family, friends, writing, and traveling. However, that retirement phase is not even in her current sight as she continues along her career path.

Part VIII

The Survey

As part of my research for this book, I orchestrated a survey of the entertainment lighting industry. I prepared a series of questions based upon the topics I planned to write about. Then I sent the survey link to over 200 targeted professionals (mostly in the USA) as well as a similar version that I opened to the wide world of social media. Overall, I received just under 1,000 responses.

As I combed through the data, I realized that it was very interesting to me and thus likely interesting to you as well. Therefore, I have included this section to share the results. What follows are charts, assessments, and quotes gathered through my research exercise.

44 Respondent Backgrounds

First I asked the survey takers to let me know how long they have worked in the entertainment lighting field. The answers were incredibly scattered from a few months to over 50 years.

The next question centered around their aspirations for career growth. The responses were widely varied, but most have a desire to continue growing their career. A respectable number are happy right where they are, while a few (the older generation) are looking towards retiring soon. I was surprised by the seven percent who said they are planning to leave our industry, and the five percent that are only motivated to make more money. However hope was not lost, as small percentages are looking for improved work/life balance, starting their own lighting company, or teaching others.

Specifically for the purposes of writing this book, I asked what career advice they desire. As expected, business and networking (socializing) skills topped the list.

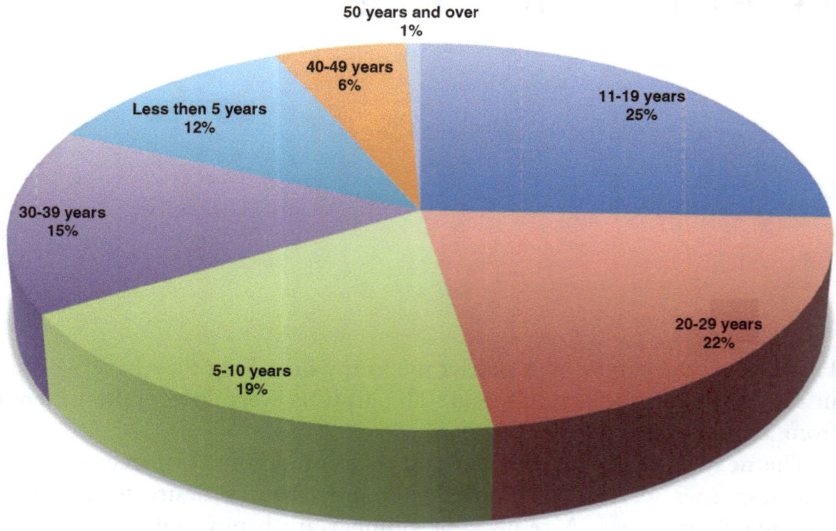

Figure 44.1 How Long Have You Been Working in Lighting?

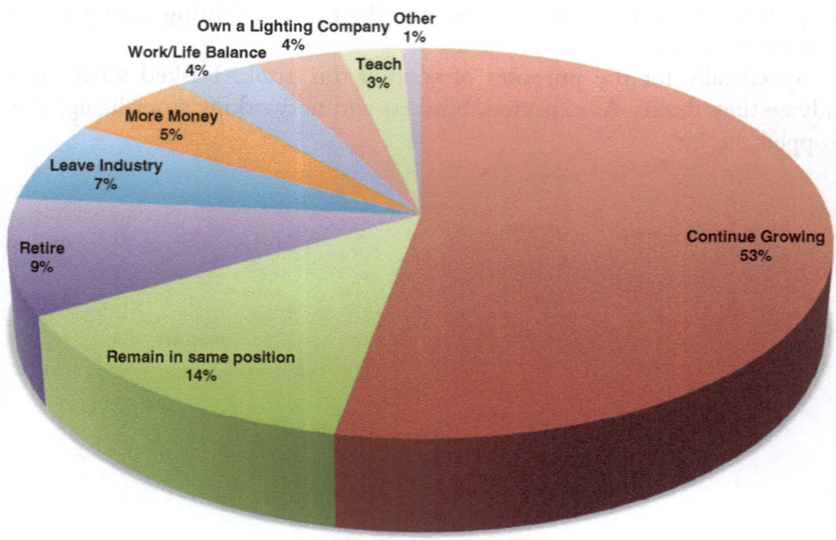

Figure 44.2 What Aspirations Do You Have for Career Growth?

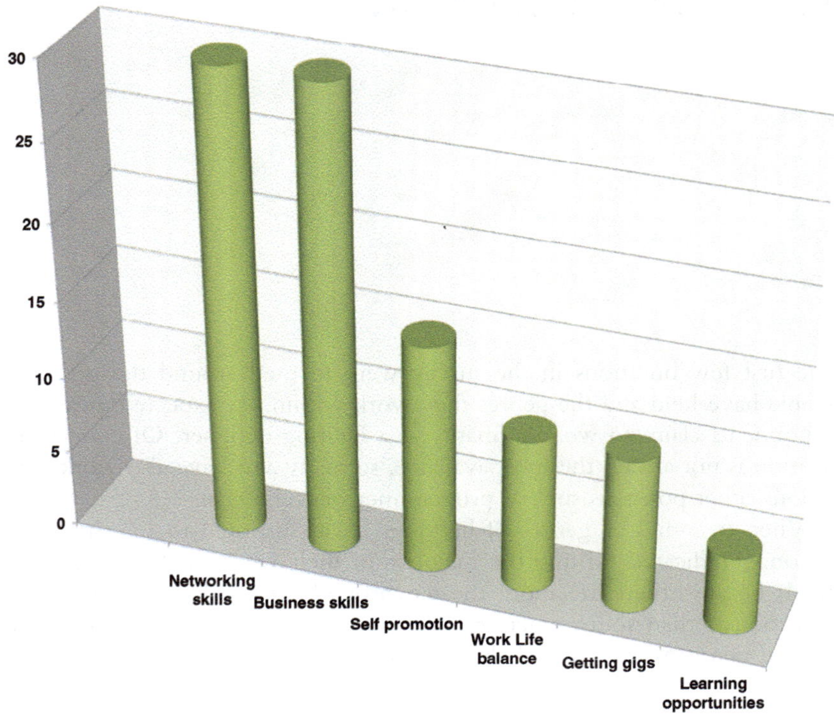

Figure 44.3 What Advice Do You Wish Someone Would Give to You to Help Your Career?

45 Employment Status

The first few questions in the survey were focused around the positions people have held and the genres they work within. As expected, nearly all respondents claim to work primarily as a lighting designer. Of course this position is not always the most available, so many also have also worked in various other positions such as programmer or technician.

When it comes to genres of lighting, I was surprised to see that most responses indicated earning the majority of their work in theater. I think this is because there are more theaters around the world (local, regional, educational) than touring acts or corporate events. However, those two popular genres were not far behind either.

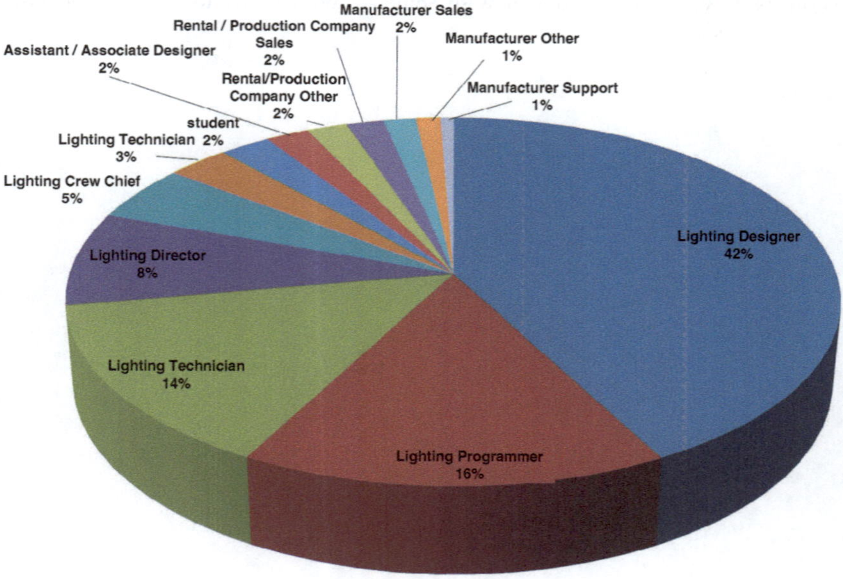

Figure 45.1 What Lighting Position Do You Mostly Work?

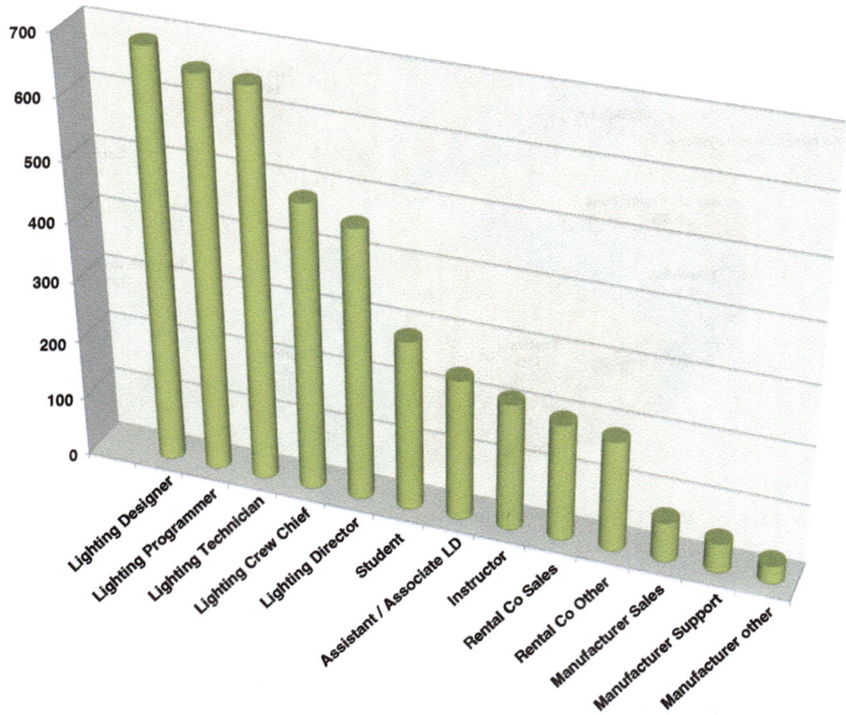

Figure 45.2 What Lighting Positions Have You Worked?

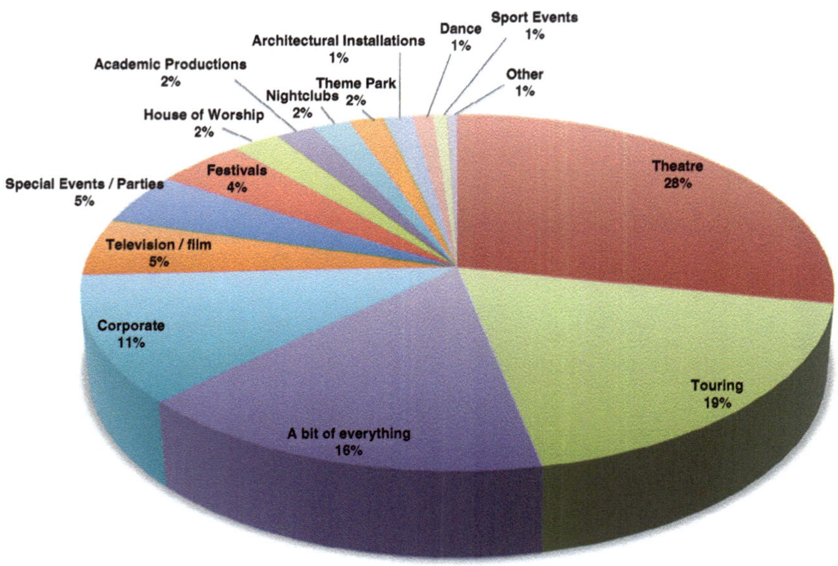

Figure 45.3 In What Genre Do You Mostly Work?

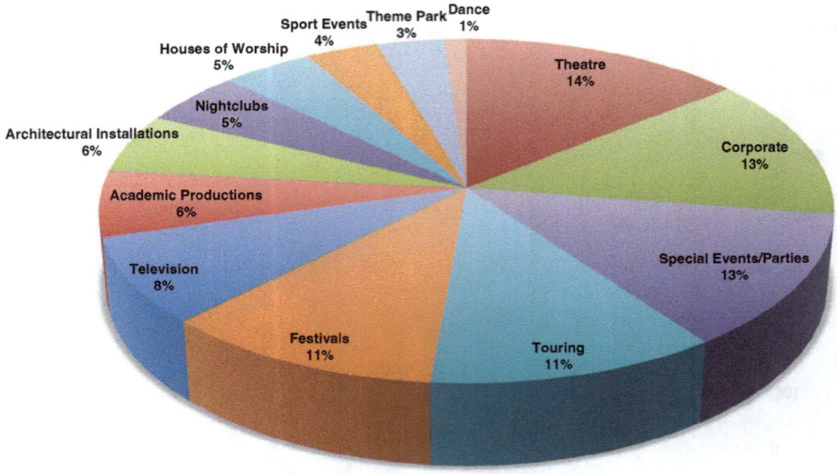

Figure 45.4 In What Genres Have You Worked?

Our industry is rather well rounded in the fact that we often work in different genres. The survey results certainly reflect that as being the case. When asked to list all the genres they have worked within, the respondents provided a broad range.

46 Down to Business

A series of questions were posed regarding employment status, insurance, taxes, and the general business of lighting. I would have thought that more people had organized their work life into a formal business, but it seems instead most are working as self-employed.

Of course, the very nature of taxes and insurance differs greatly from country to country. In the United States, we don't have provided medical

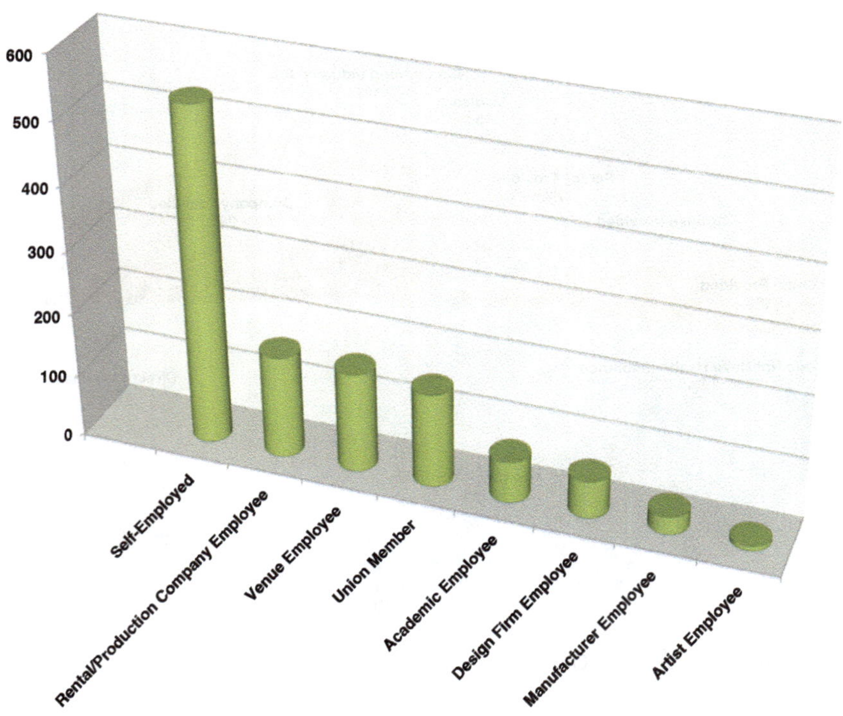

Figure 46.1 How Are You Employed?

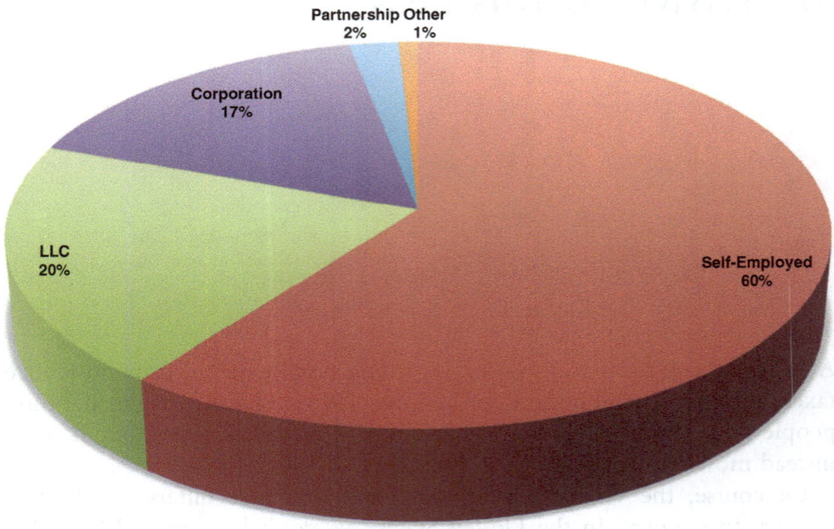

Figure 46.2 How Do You Work as Self-Employed?

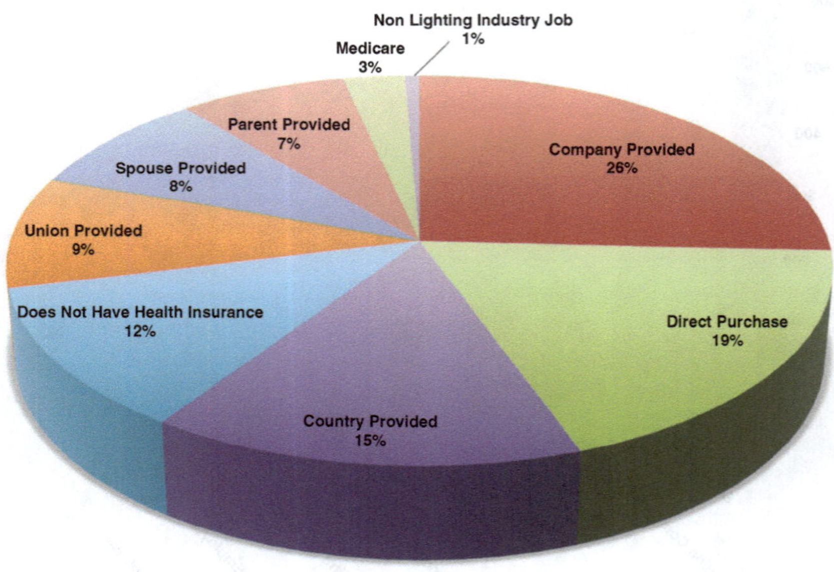

Figure 46.3 How Do You Get Health Insurance?

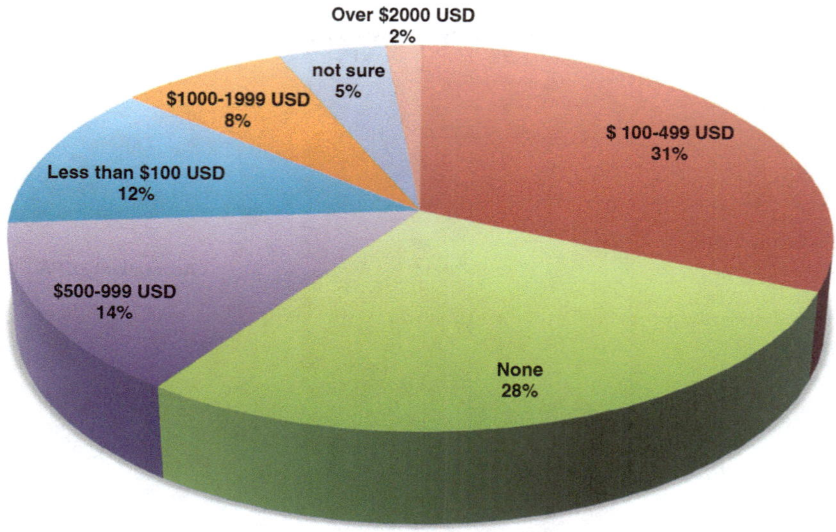

Figure 46.4 Health Insurance Cost Per Month

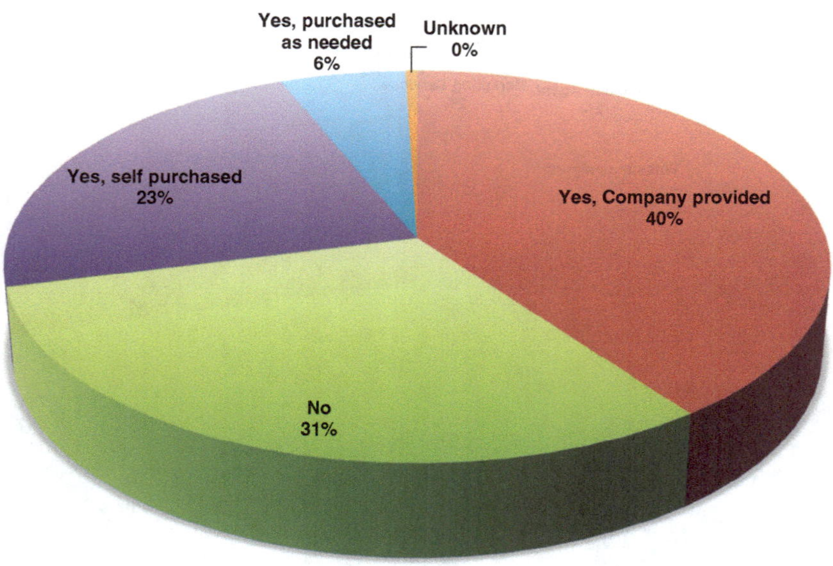

Figure 46.5 Do You Have Liability or Worker's Comp Insurance?

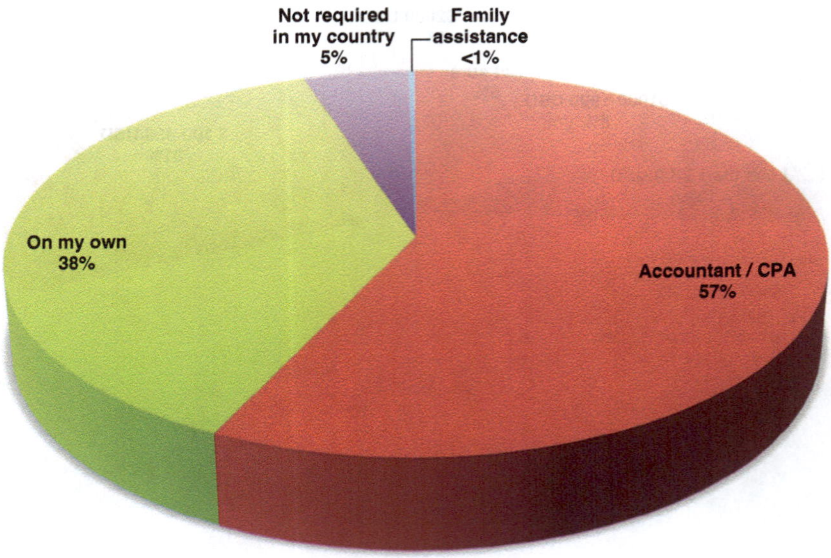

Figure 46.6 How Do You Complete Your Taxes?

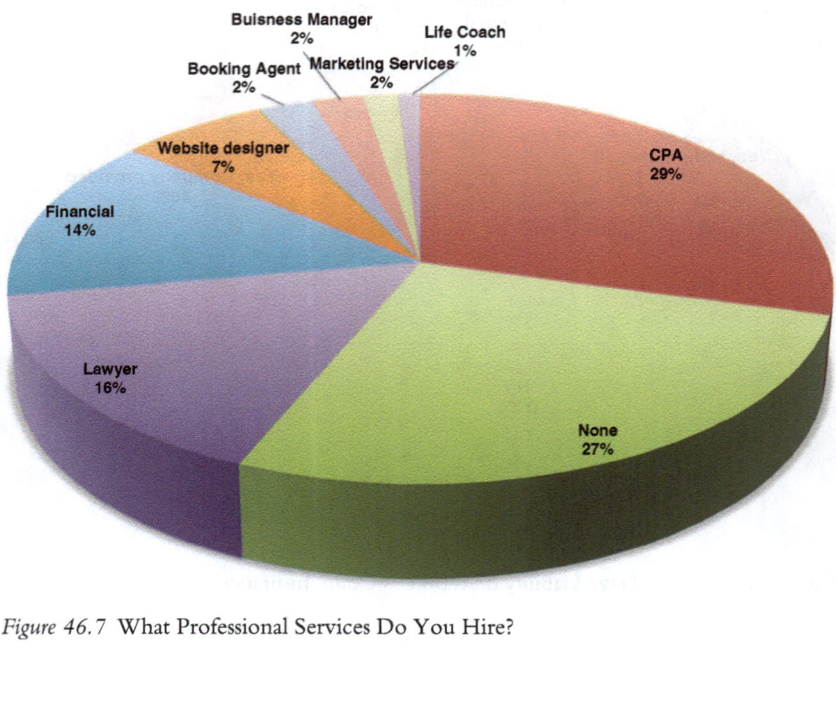

Figure 46.7 What Professional Services Do You Hire?

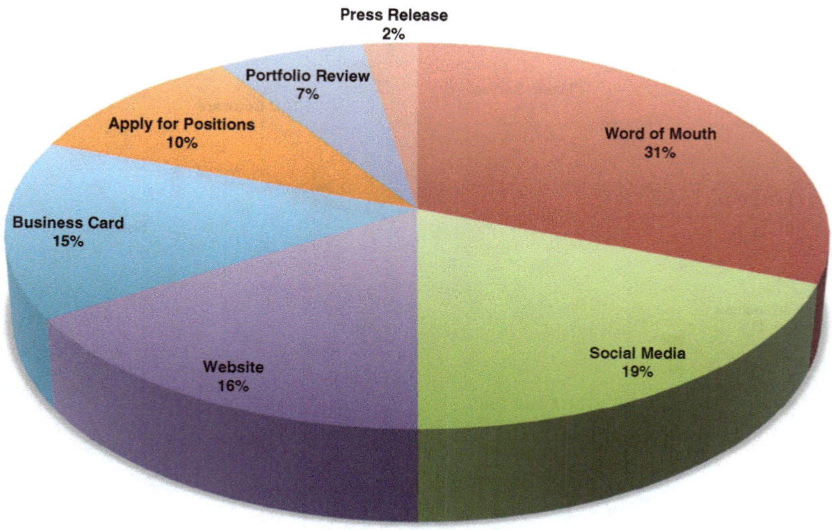

Figure 46.8 How Do You Promote Yourself in Your Career?

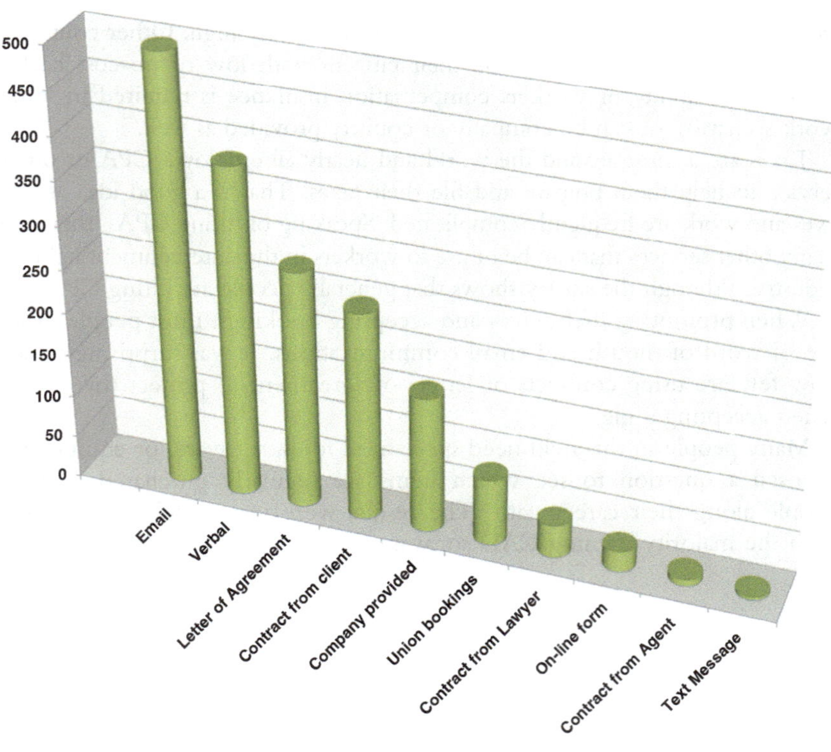

Figure 46.9 How Do You Confirm Your Bookings?

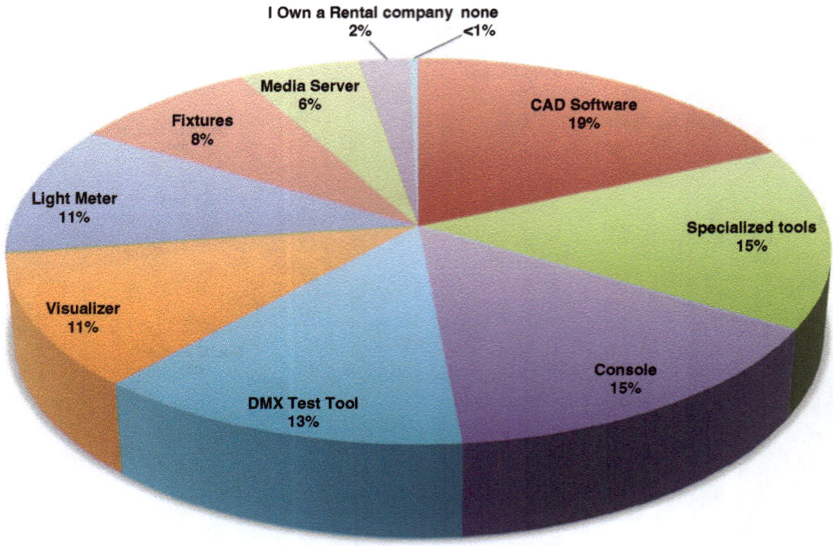

Figure 46.10 What Lighting Gear Do You Own?

care and thus the costs for health insurance are usually high. Other countries tax at a higher rate, but provide their citizens with low or no-cost health insurance. Liability or workers compensation insurance is required in some work scenarios, or can be company or country provided as well.

Taxes are a must around the world and nearly all employ a CPA or other service to help them prepare and file their taxes. That is a good idea as our lives and work are frequently complicated. Speaking of hiring CPAs, there are many other services that can be of use to workers in the entertainment lighting industry, although the survey shows that generally people are hiring CPAs.

When promoting themselves and accepting bookings, most people make use of word of mouth and email communications. It was surprising to see how few are using contracts or letters of agreement to protect themselves when accepting a gig.

Many people in this field need specialized tools, software, or equipment. I posed a question to see which items are regularly purchased to help people along their career path. The results were very much as expected, with the majority owning CAD software.

47 Shared Advice

The survey also asked respondents a more personal question regarding advice they wish to share with others. The results were fascinating, with many centering around similar themes. What follows are some of the best advice quotes shared from our wonderful industry colleagues, broken down into four main categories.

Attitude

- Talk less; smile more.
- Believe in yourself and your abilities, but always listen and work hard.
- Own everything you do, from mistakes to successes.
- Be nice to people, or, at the very least, don't be a jerk.
- Be the sort of person you'd like to work with.
- Keep your head down, keep your work ethic up, keep your mouth shut, and do what you do best.
- Two ears, one mouth.

Design

- If you want to design, design.
- Design should never be about one's self or how you can boost your ego. It is about helping to create a memorable aesthetic experience.
- Lighting design is not about the manipulation of light, but rather an orchestration in shadow.
- Pick three colors, and then use two.
- Light the performers first. Remember darkness is as powerful as light.
- There is more out there than theater to design.
- The lighting is the easy part; it's the politics that's difficult.

Career

- You have to learn to use a shovel before you can drive the big machine.
- Always be a go-getter. Educate yourself, seek out opportunities to take classes, in order to learn and meet people. Don't sit back and wait for things to fall into your lap, or to be paid to learn something.
- Don't make career decisions based solely on financial gain. What you choose to do matters more than what you get paid to do it.
- There will always be someone ready to take your place, so don't be complacent.
- If you are a female in the industry, it will be one of the hardest things you will have to do. You have to work twice as hard to be given the same respect as a male walking into the room.
- Some jobs are for rent, some are for art, some are stepping-stones for growth.
- Seek out mentors.

Life

- What makes you happy is the most important. You are the one who has to look at yourself in the mirror every day, so don't make your life-changing decisions based off what your mentor, colleagues, or peers in the field want you to do. Do what makes you the happiest, because that is what will allow you to enjoy life.
- If you are an enjoyable person to work with, you will get asked to the bar after work. If you get asked, say yes, and don't get drunk.
- You can't judge your own success by measuring it to someone else's. Your path will be unique, but it will not be straight or easy.
- Be exceptional, gain the knowledge, take the training, and get the experience to understand the real and potential risks of any workplace.
- Do what YOU want; it's your life.

48 Wrapping It Up

While only a small percentage of the responders said they were close to retirement, many were already saving money for their future. The methods of planning vary, with a majority simply saving money or investing in an IRA.

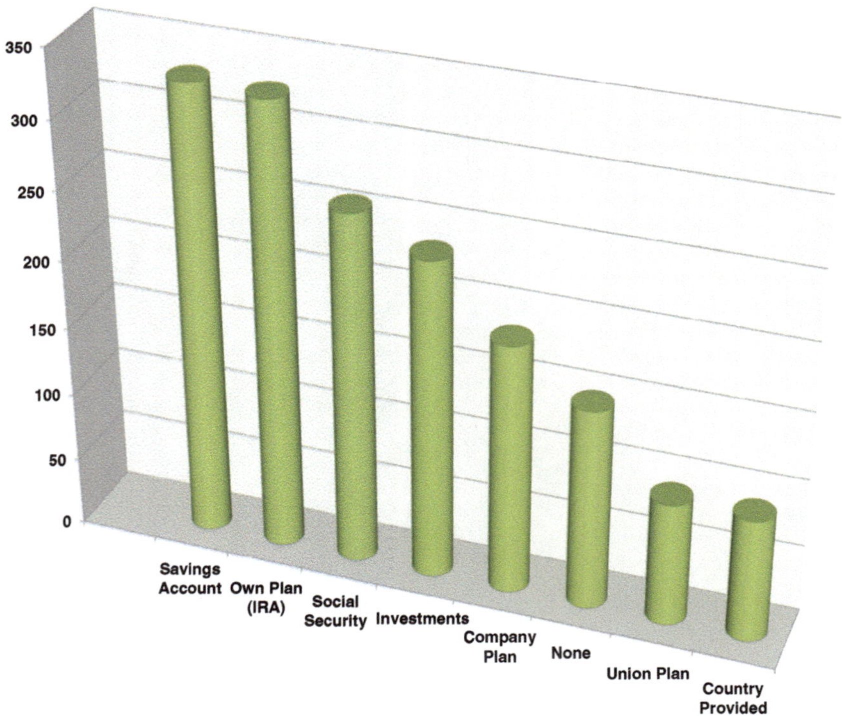

Figure 48.1 How Have You Prepared for Retirement?

A Simple Plan

Respondents were also very open to explaining their guidelines for a successful career. As I pored over the answers, it was clear that there were three overwhelming points shared by a vast majority of the survey respondents. I also agree that these are the key tenements to a successful career in the entertainment lighting industry:

1. Show up on time and never be late, no matter your position or stature.
2. Work hard, never stop learning, and always have fun.
3. Be nice and treat others with respect.

This survey occurred in early 2019 and it clearly confirmed that the above points have stood the test of time. While they can hold true to most industries, they are especially important for success in the entertainment lighting field. May you go forth and have a wonderful career.

Index